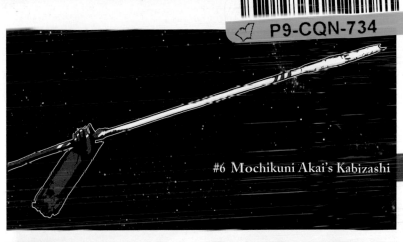

#6 Mochikuni Akai's Kabizashi

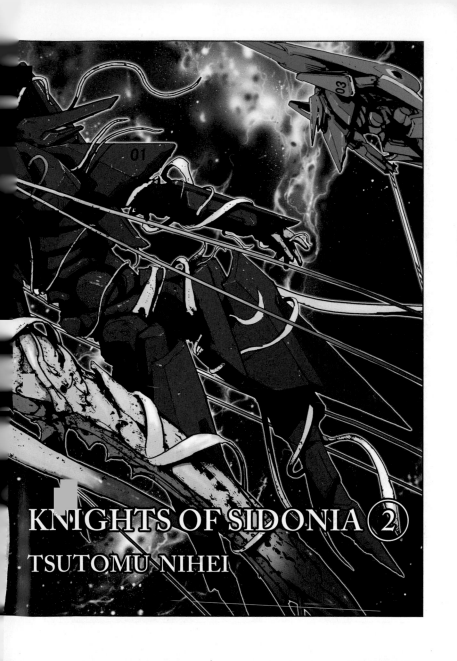

KNIGHTS OF SIDONIA ②

TSUTOMU NIHEI

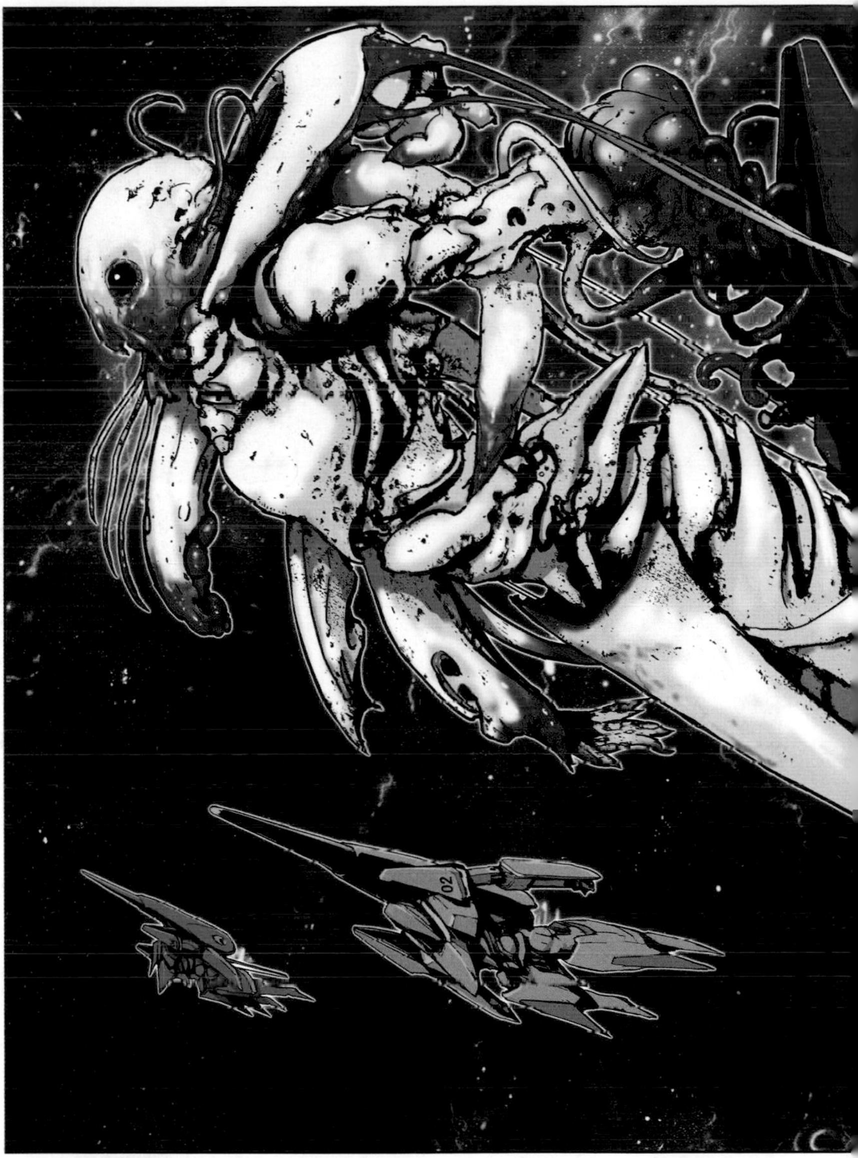

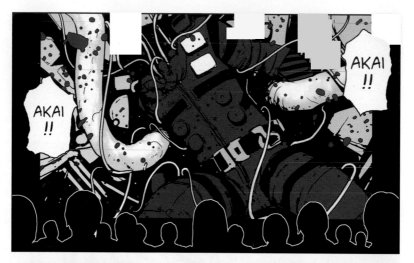

AKAI !!

AKAI !!

WHAT THE ...

HEY! WHAT WAS ...?!

BZZT

END OF BROADCAST

I-IT CAN'T BE...

MR. AKAI'S ...

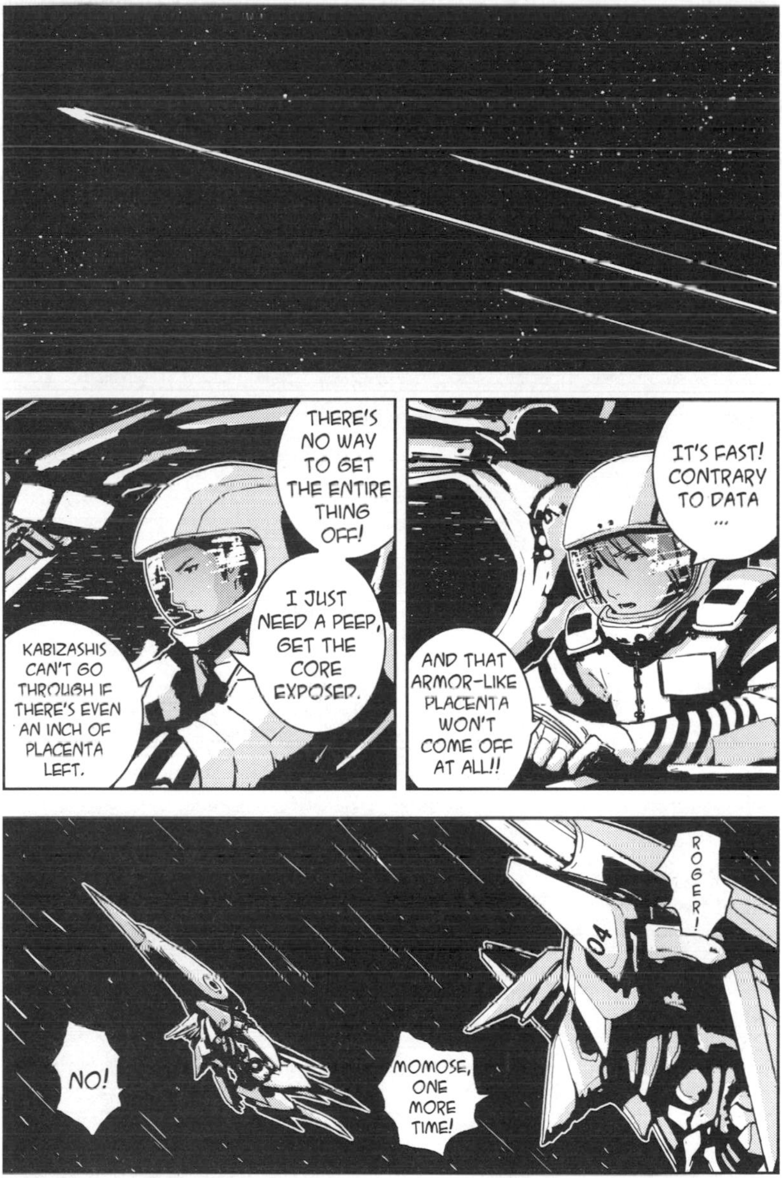
IT'S FAST! CONTRARY TO DATA ...
AND THAT ARMOR-LIKE PLACENTA WON'T COME OFF AT ALL!!
THERE'S NO WAY TO GET THE ENTIRE THING OFF!
I JUST NEED A PEEP, GET THE CORE EXPOSED.
KABIZASHIS CAN'T GO THROUGH IF THERE'S EVEN AN INCH OF PLACENTA LEFT.
ROGER!
04
MOMOSE, ONE MORE TIME!
NO!

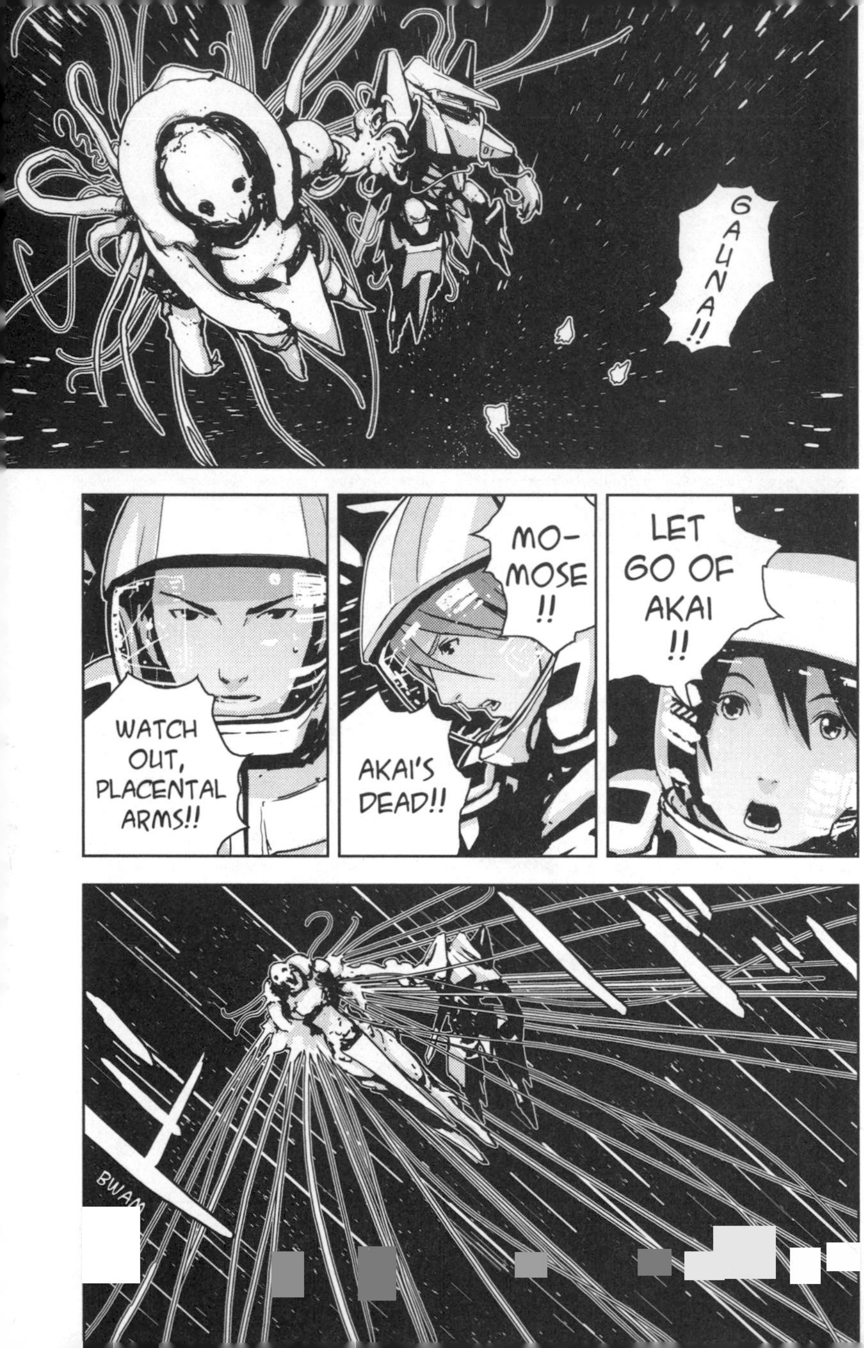
GAUNA!!
LET GO OF AKAI !!
MO-MOSE !!
AKAI'S DEAD!!
WATCH OUT, PLACENTAL ARMS!!
BWAM

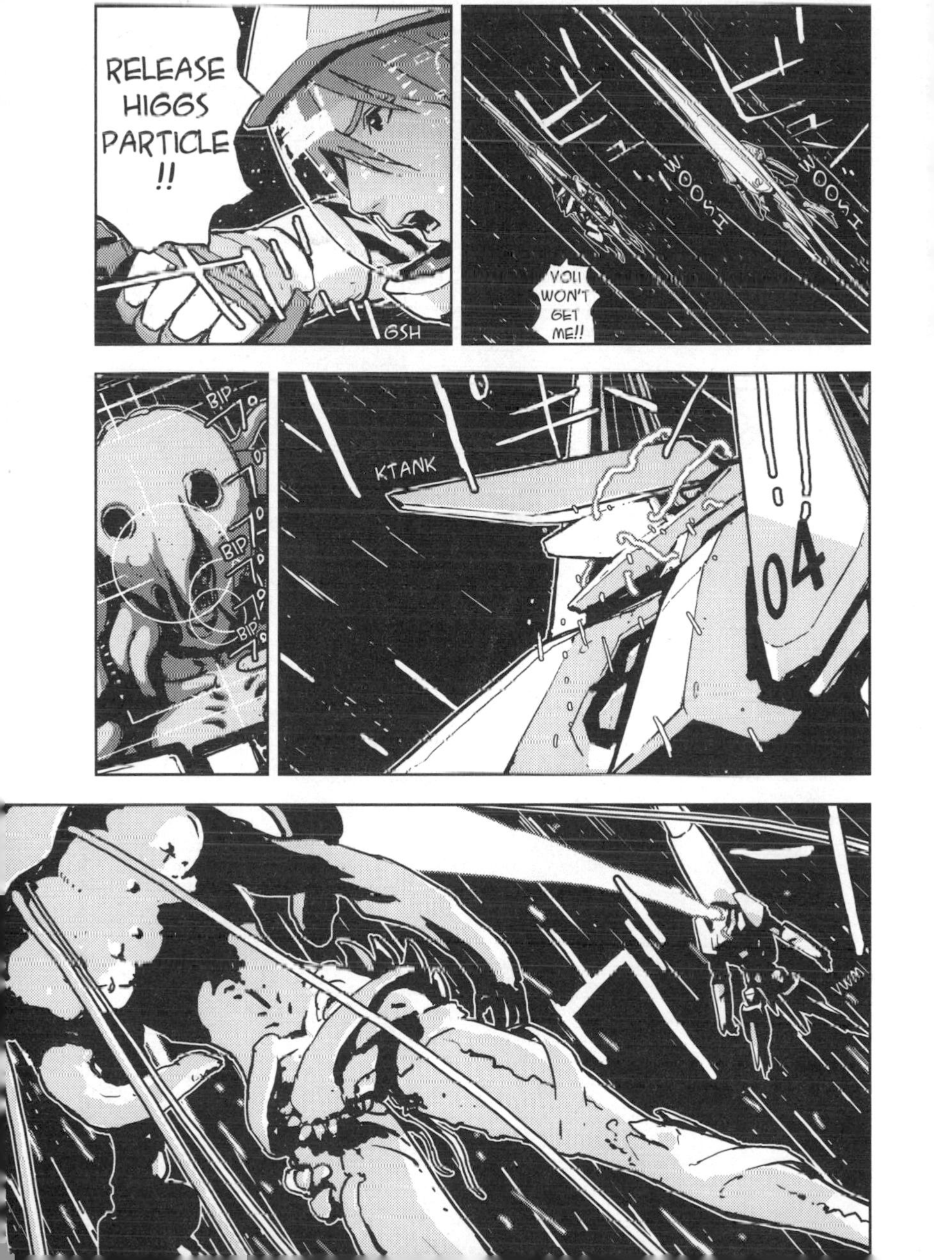
YOU WON'T GET ME!!
RELEASE HIGGS PARTICLE!!
GSH
KTANK
BIP
BIP
BIP

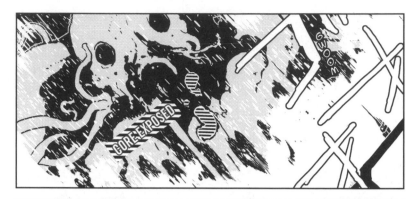

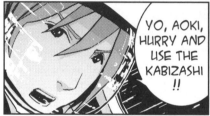

GWOOM
CORE EXPOSED
YEAH!! THERE!
WE CAN SEE THE GAUNA'S CORE!
YO, AOKI, HURRY AND USE THE KABIZASHI !!
!!
AOKI !!
03

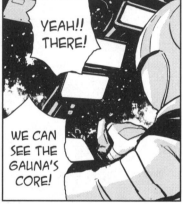

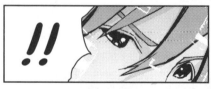

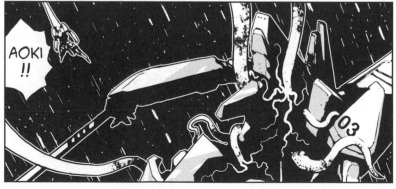

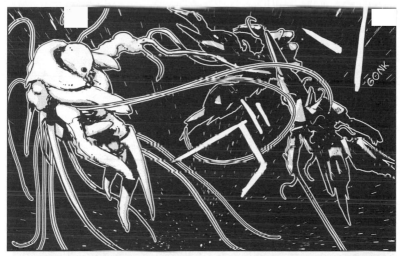

GONK

DVOOM

AKAI AND AOKI UNITS HEAVILY DAMAGED !!

BOTH KABIZASHIS ARE ADRIFT !!

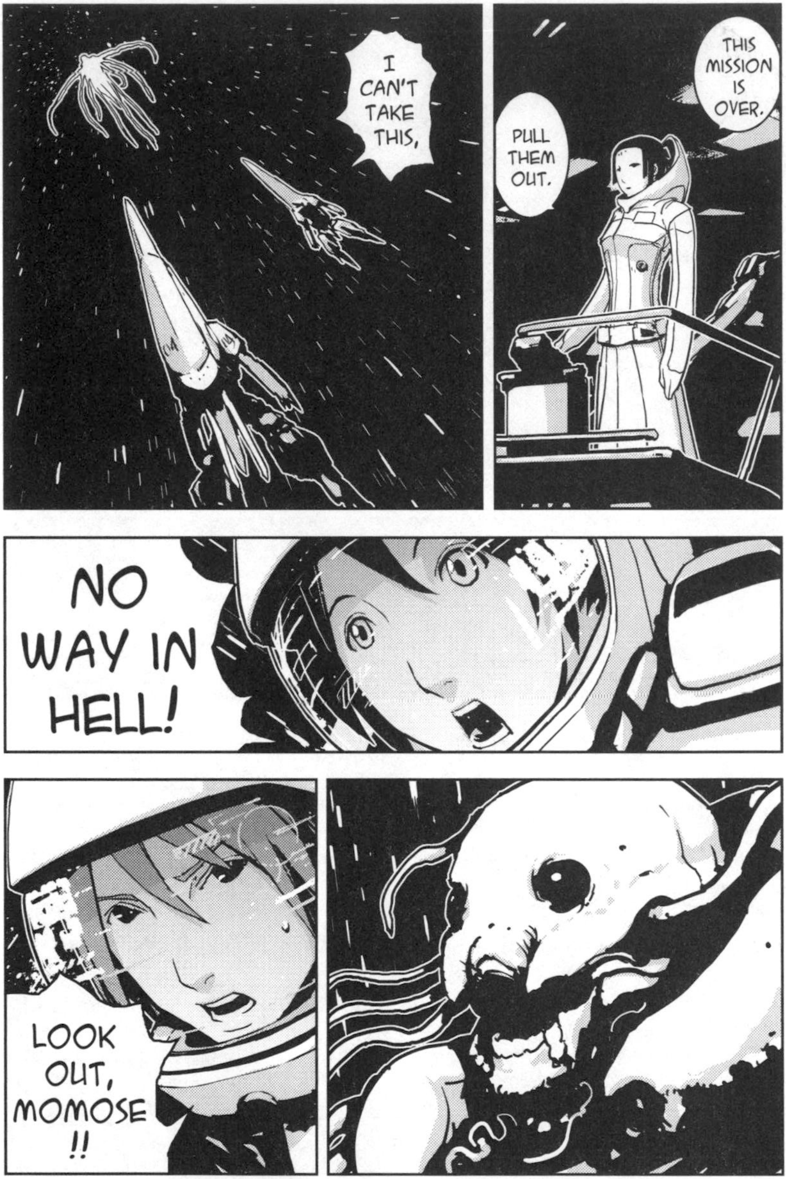
THIS MISSION IS OVER.
PULL THEM OUT.
I CAN'T TAKE THIS,
NO WAY IN HELL!
LOOK OUT, MOMOSE!!

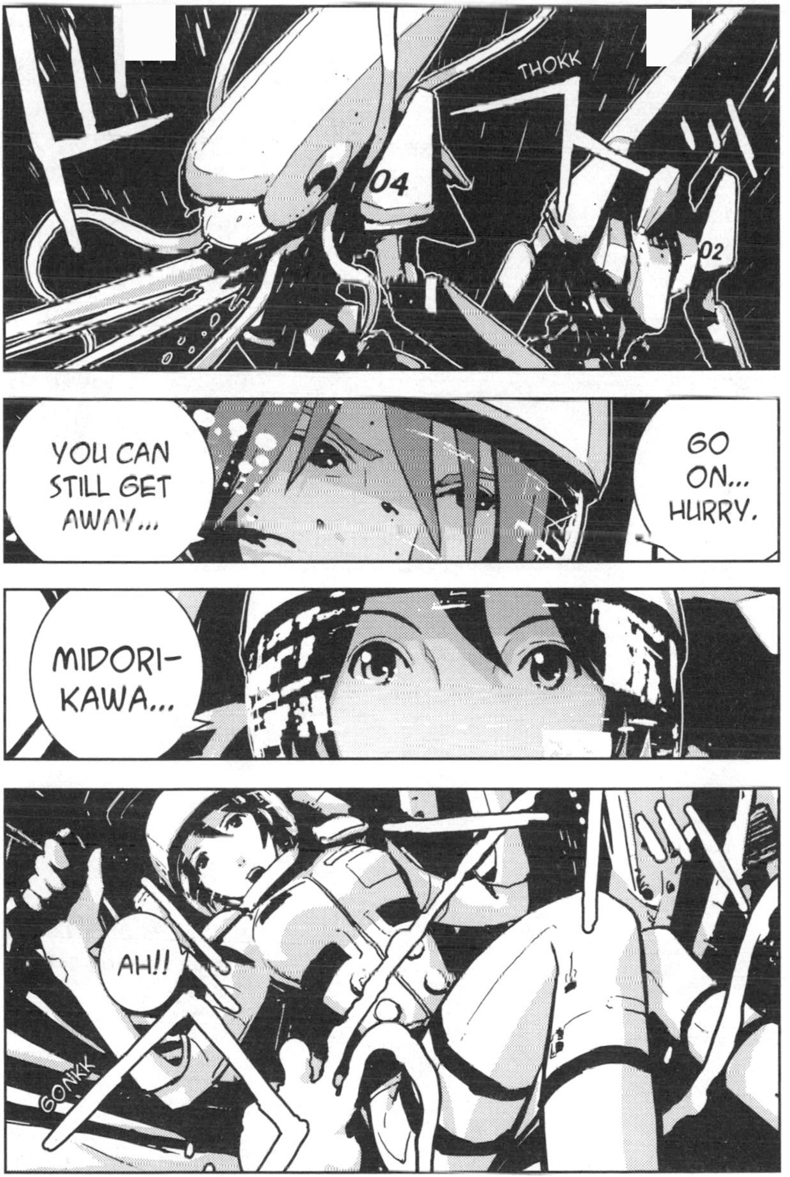
THOKK
04
02
GO ON... HURRY.
YOU CAN STILL GET AWAY...
MIDORI-KAWA...
AH!!
GONKK

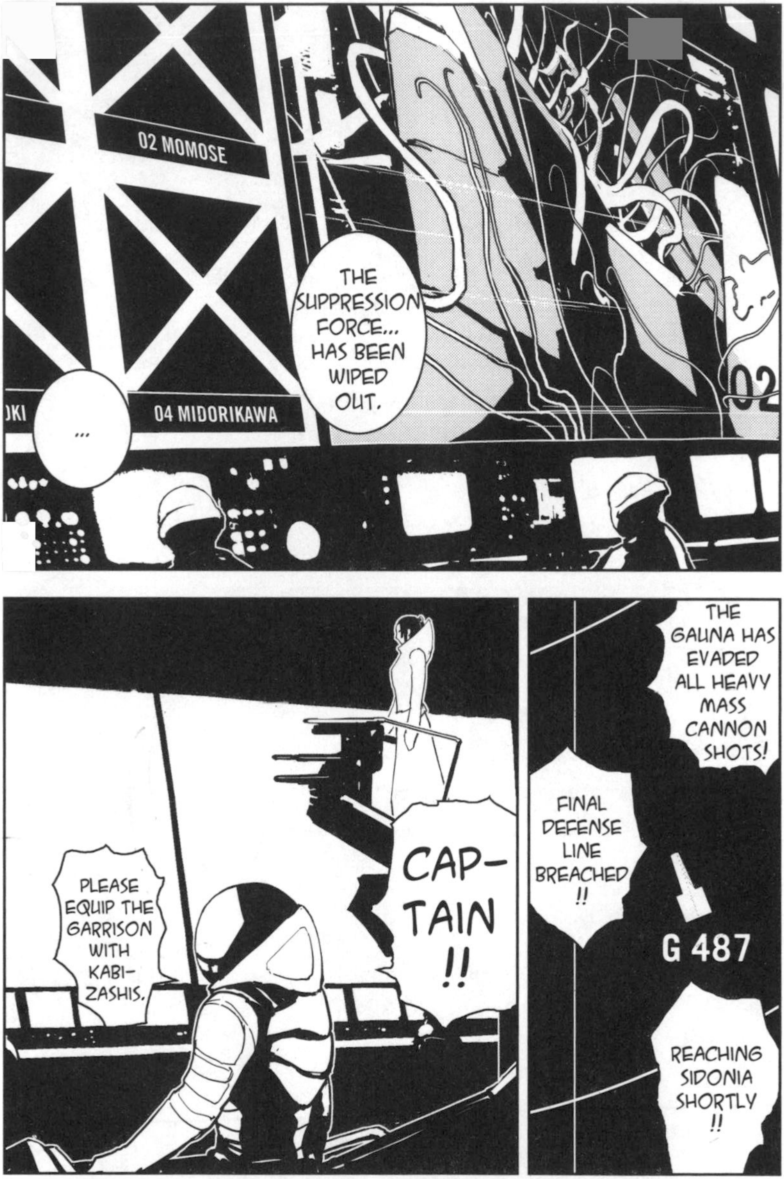
02
THE SUPPRESSION FORCE... HAS BEEN WIPED OUT.
02 MOMOSE
04 MIDORIKAWA
...
THE GAUNA HAS EVADED ALL HEAVY MASS CANNON SHOTS!
FINAL DEFENSE LINE BREACHED !!
G 487
REACHING SIDONIA SHORTLY !!
CAP-TAIN !!
PLEASE EQUIP THE GARRISON WITH KABI-ZASHIS.

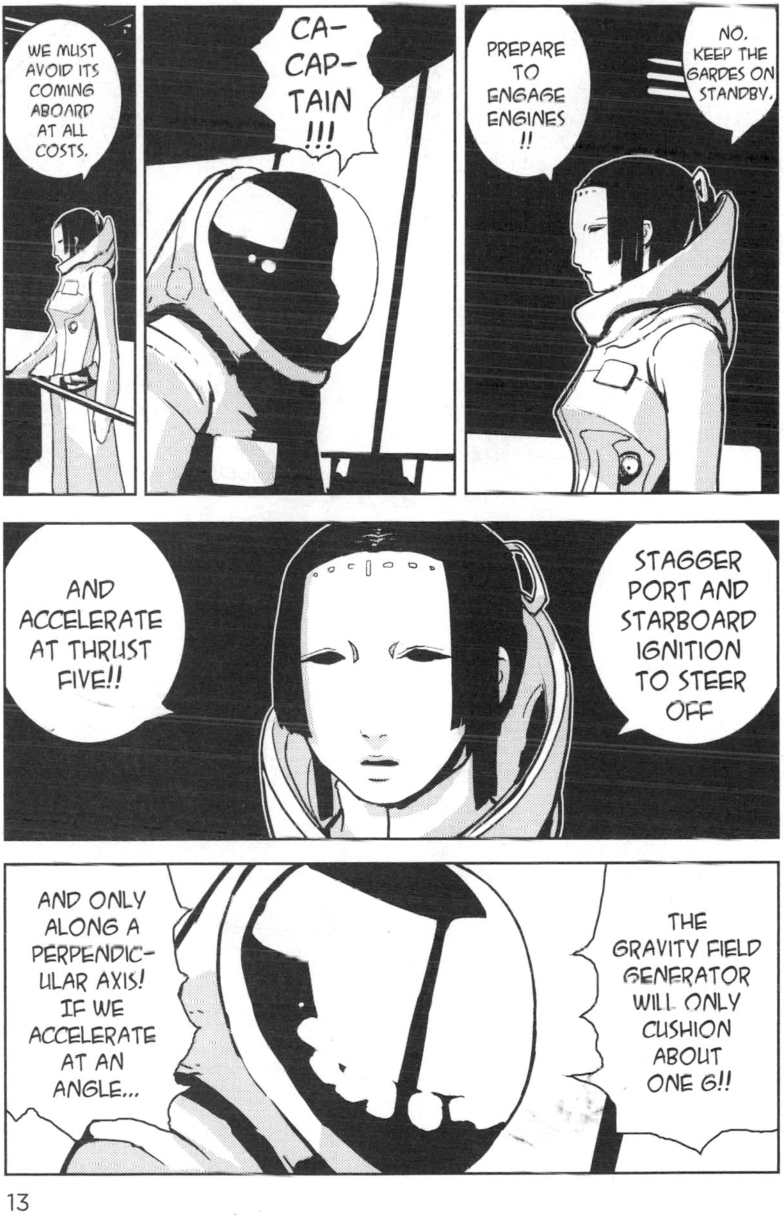
NO. KEEP THE GARDES ON STANDBY.
PREPARE TO ENGAGE ENGINES !!
CA- CAP- TAIN !!!
WE MUST AVOID ITS COMING ABOARD AT ALL COSTS.
STAGGER PORT AND STARBOARD IGNITION TO STEER OFF
AND ACCELERATE AT THRUST FIVE!!
THE GRAVITY FIELD GENERATOR WILL ONLY CUSHION ABOUT ONE G!!
AND ONLY ALONG A PERPENDIC- ULAR AXIS! IF WE ACCELERATE AT AN ANGLE...

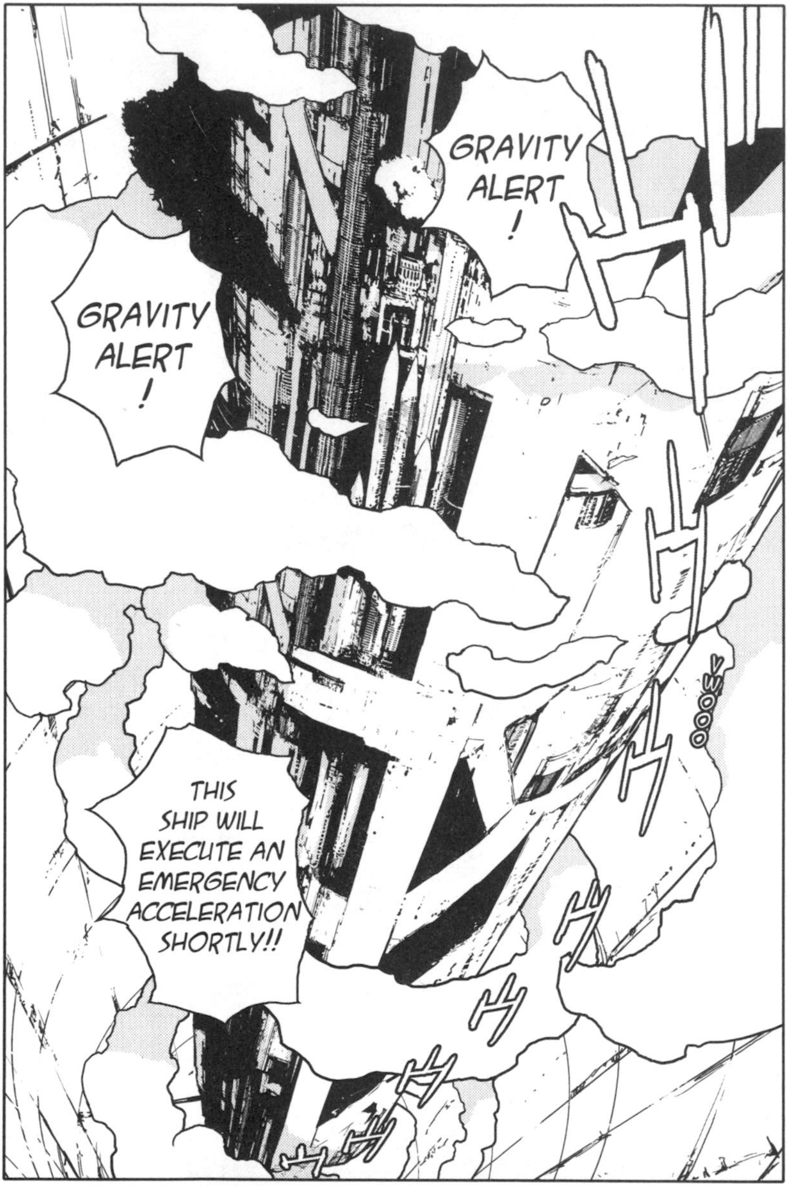
GRAVITY ALERT!
GRAVITY ALERT!
VWOOO
THIS SHIP WILL EXECUTE AN EMERGENCY ACCELERATION SHORTLY!!

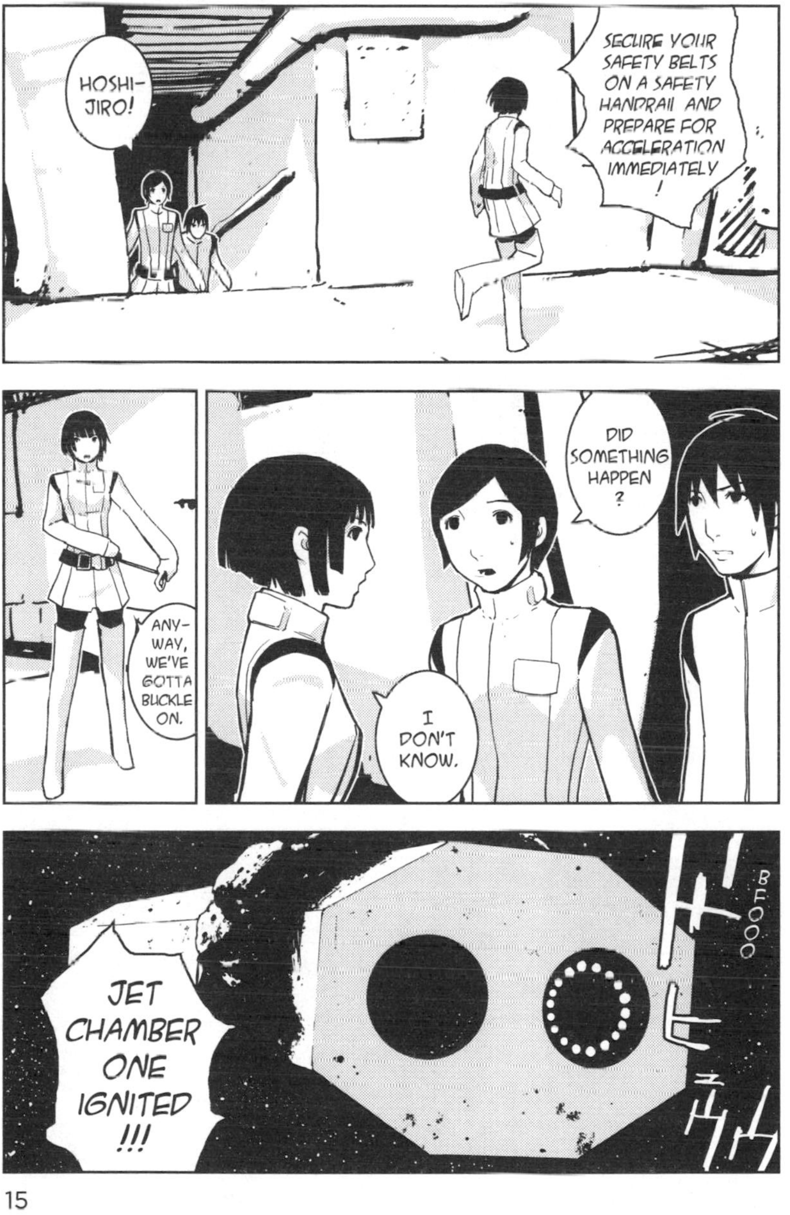
SECURE YOUR SAFETY BELTS ON A SAFETY HANDRAIL AND PREPARE FOR ACCELERATION IMMEDIATELY!
HOSHI-JIRO!
DID SOMETHING HAPPEN?
I DON'T KNOW.
ANY-WAY, WE'VE GOTTA BUCKLE ON.
BFOOO
JET CHAMBER ONE IGNITED!!!

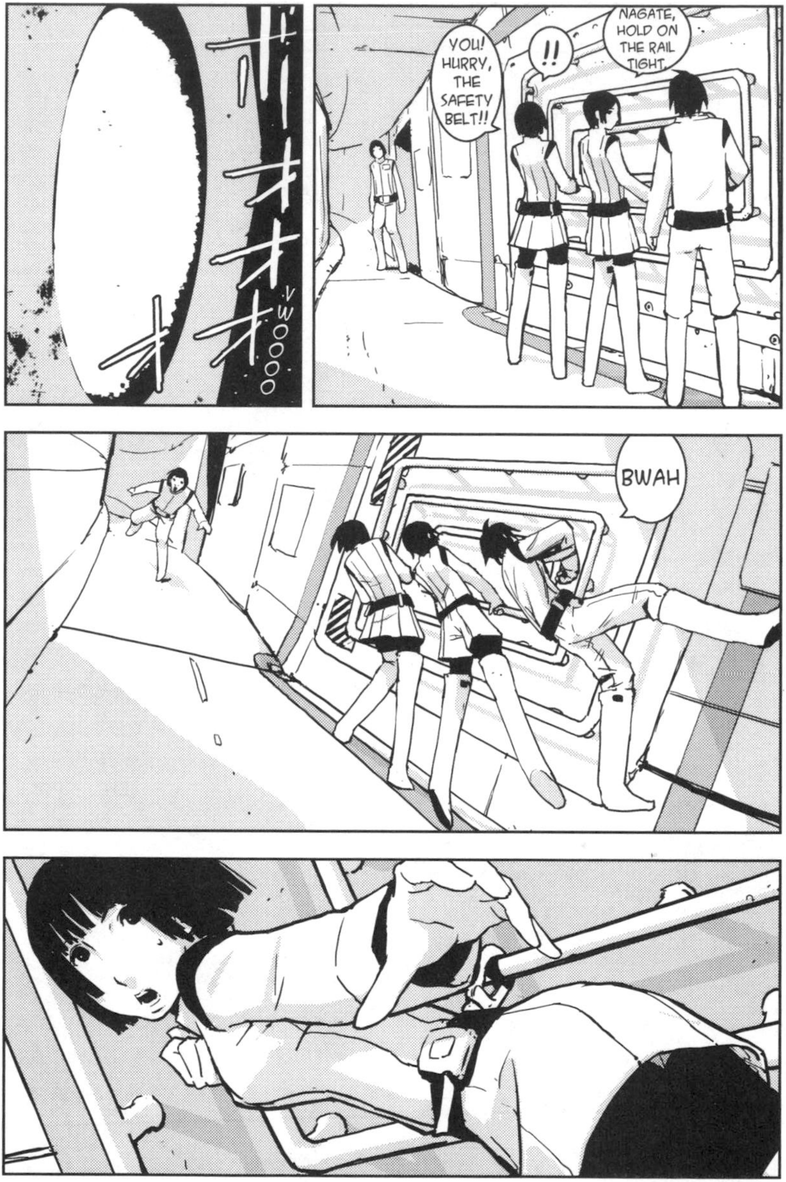
NAGATE, HOLD ON THE RAIL TIGHT.
!!
YOU! HURRY, THE SAFETY BELT!!
VWOOOO
BWAH

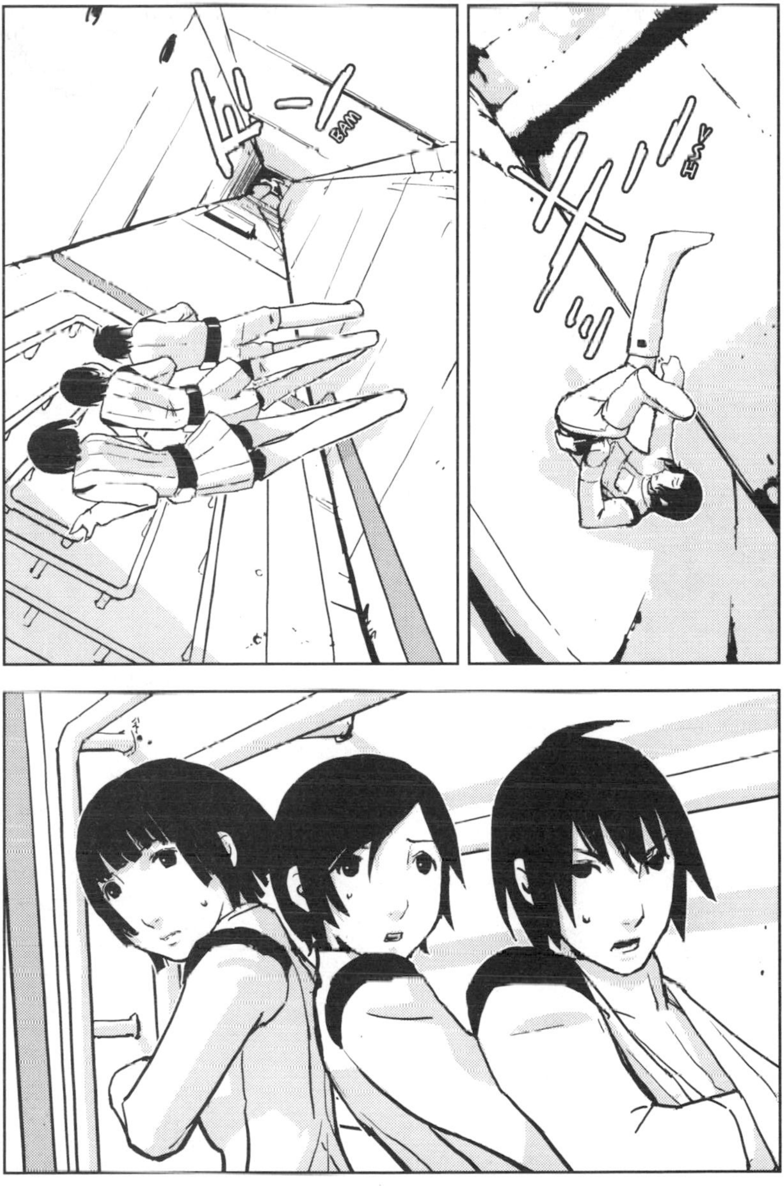
BAM

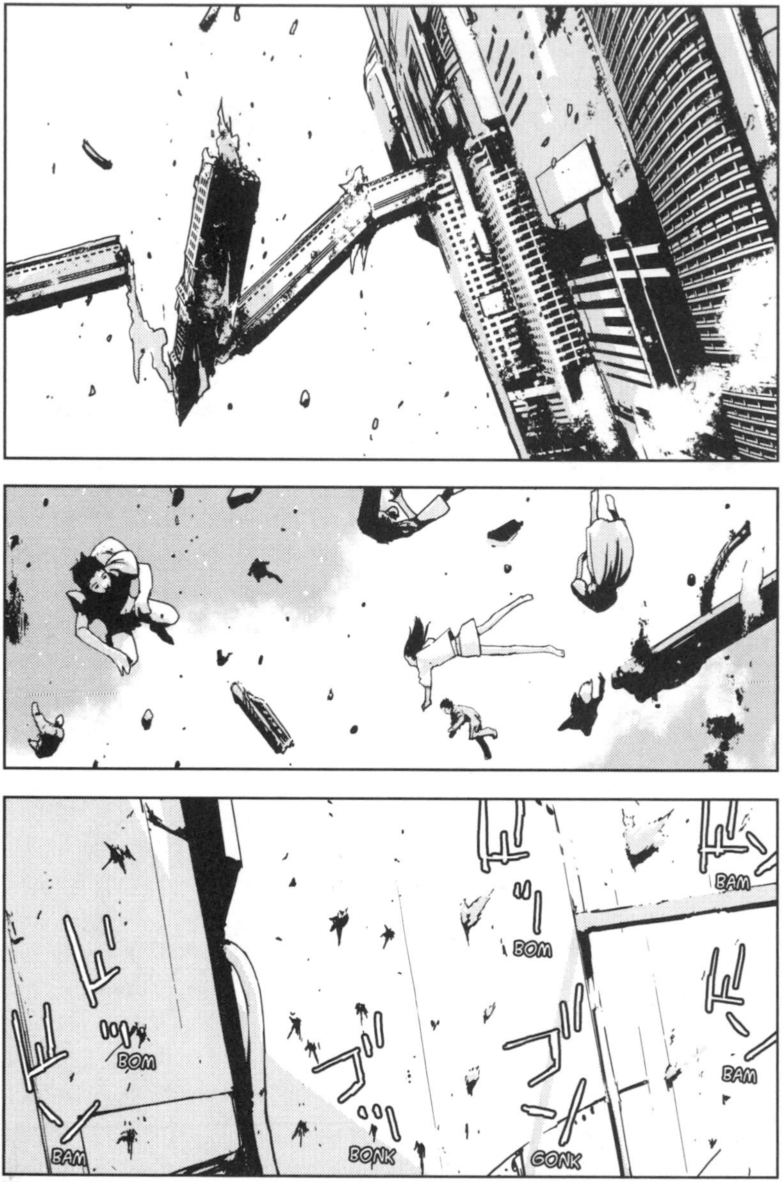
BAM
BAM
BOM
GONK
BONK
BOM
BAM

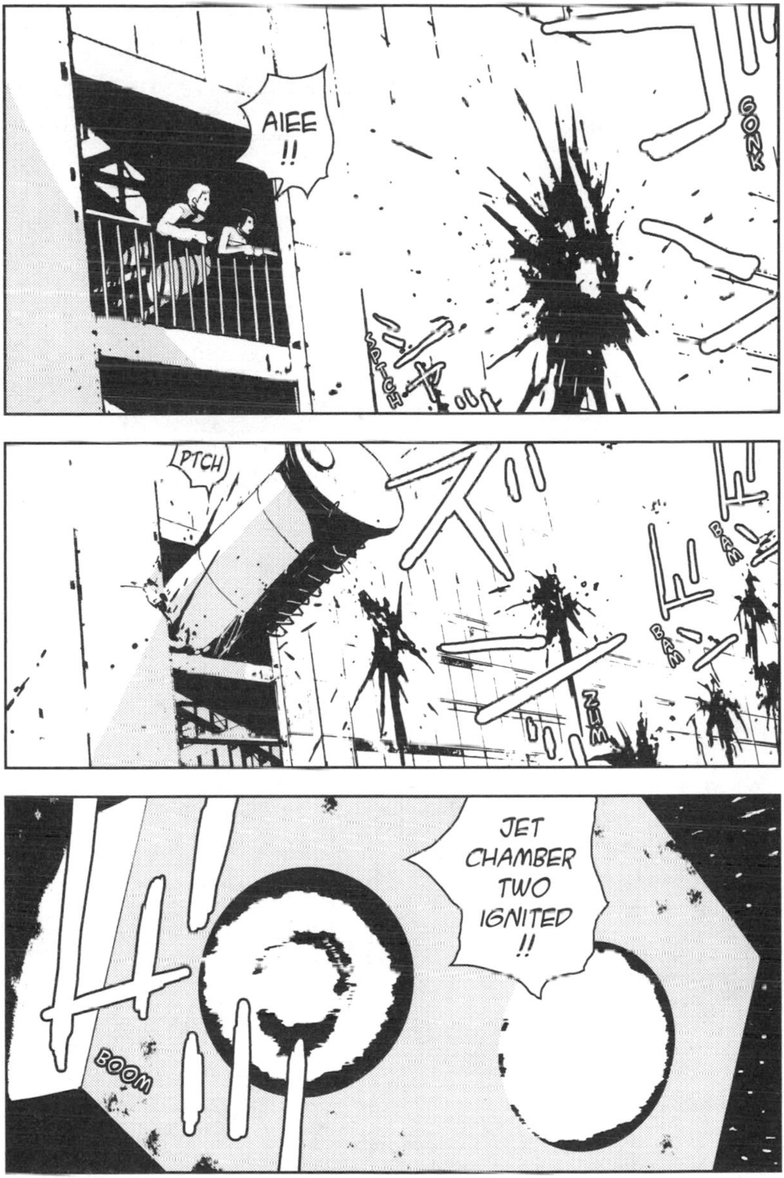
GONK
AIEE !!
SPTCH
BAM
BAM
ZUM
PTCH
JET CHAMBER TWO IGNITED !!
BOOM

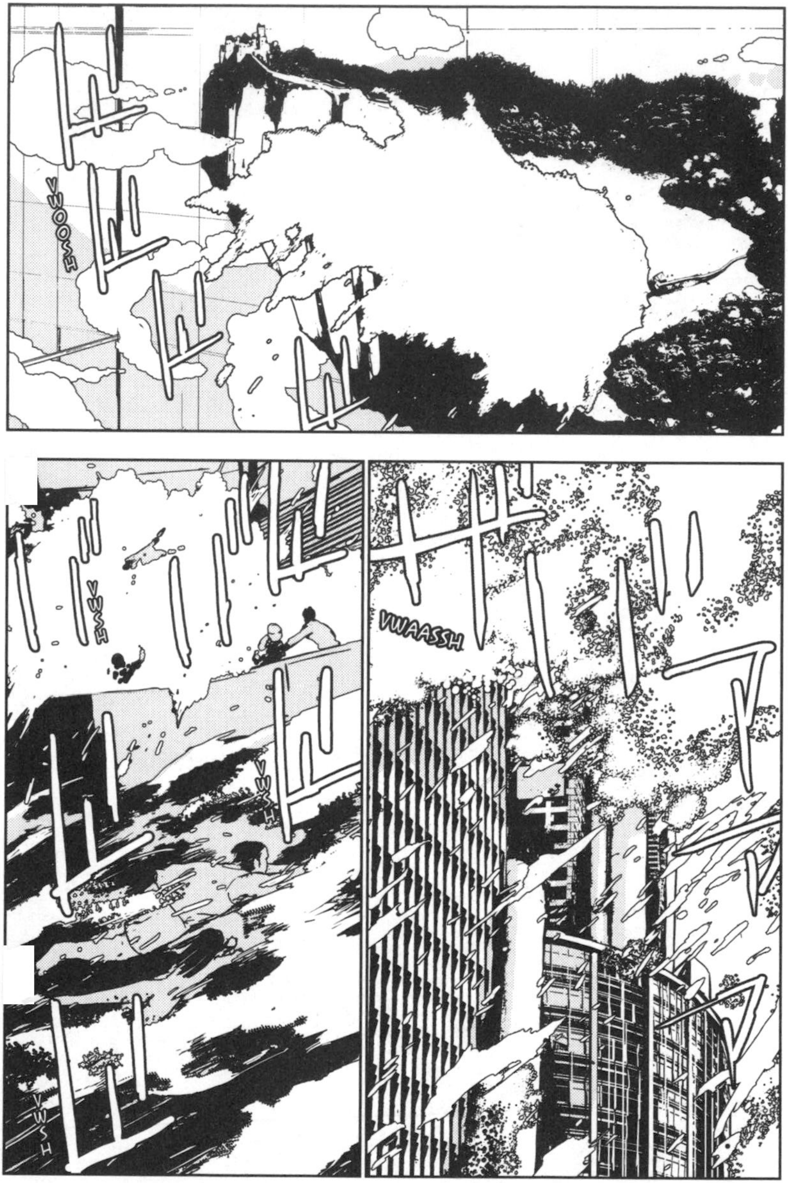
VWOOSH
VWAASSH
VWSH
VWSH
VWSH

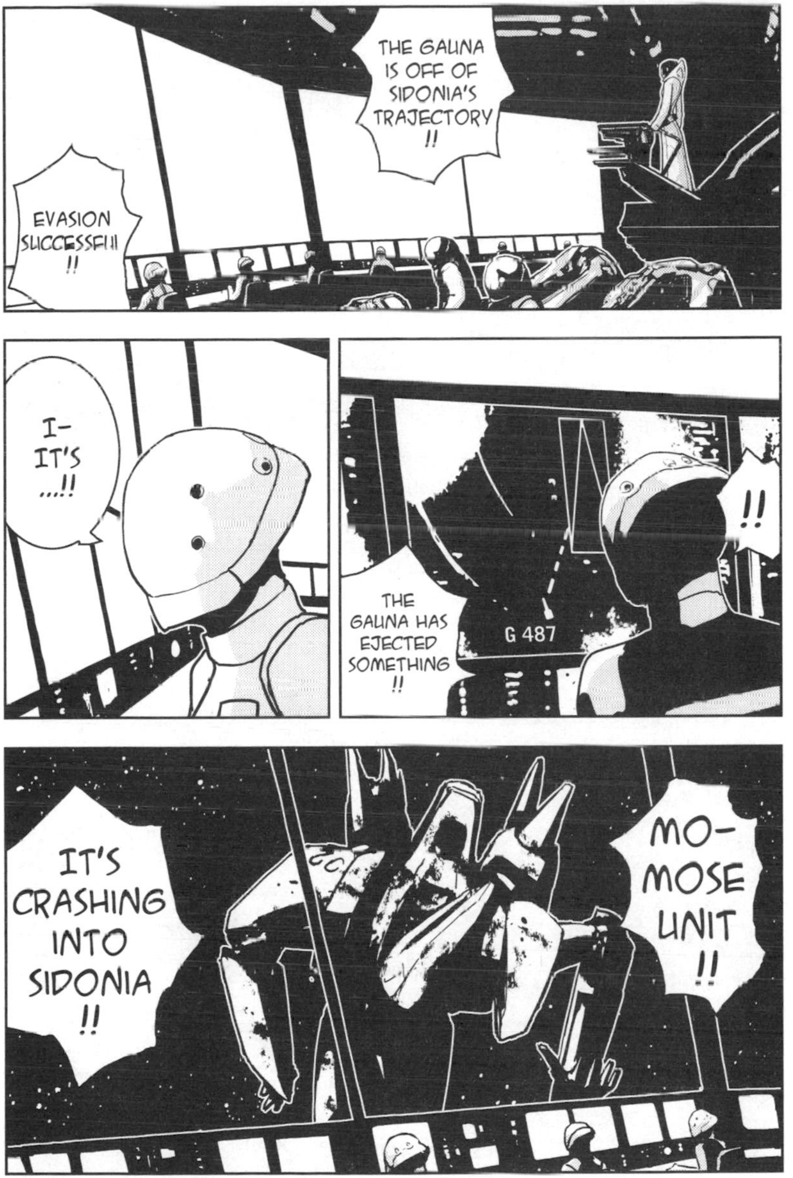
THE GAUNA IS OFF OF SIDONIA'S TRAJECTORY !!
EVASION SUCCESSFUL !!
!!
G 487
THE GAUNA HAS EJECTED SOMETHING !!
I- IT'S ...!!
MO-MOSE UNIT !!
IT'S CRASHING INTO SIDONIA !!

BOOM
HALT JETS !!
CONVERT TO INERTIAL NAVIGATION AND EQUIP THE GARRISON WITH EIGHT KABIZASHIS!

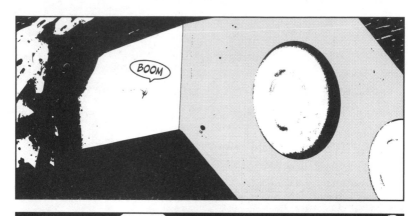

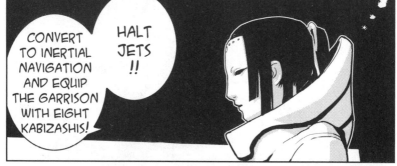

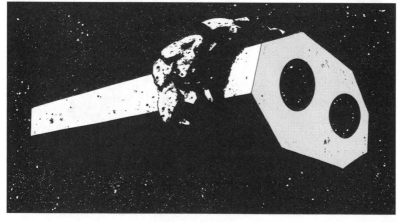

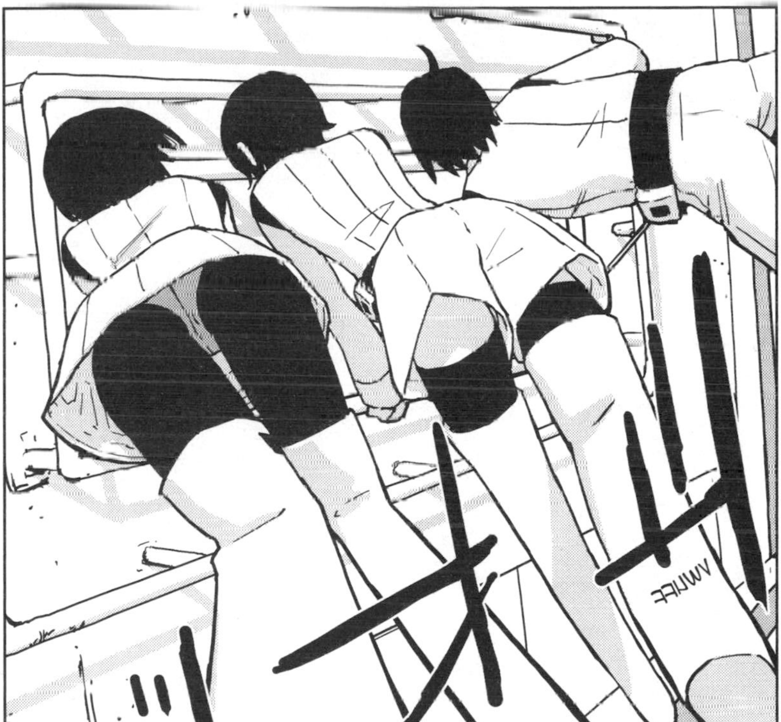

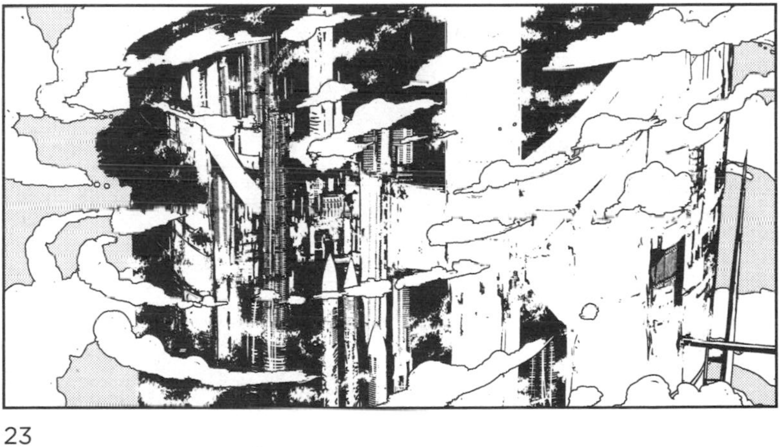

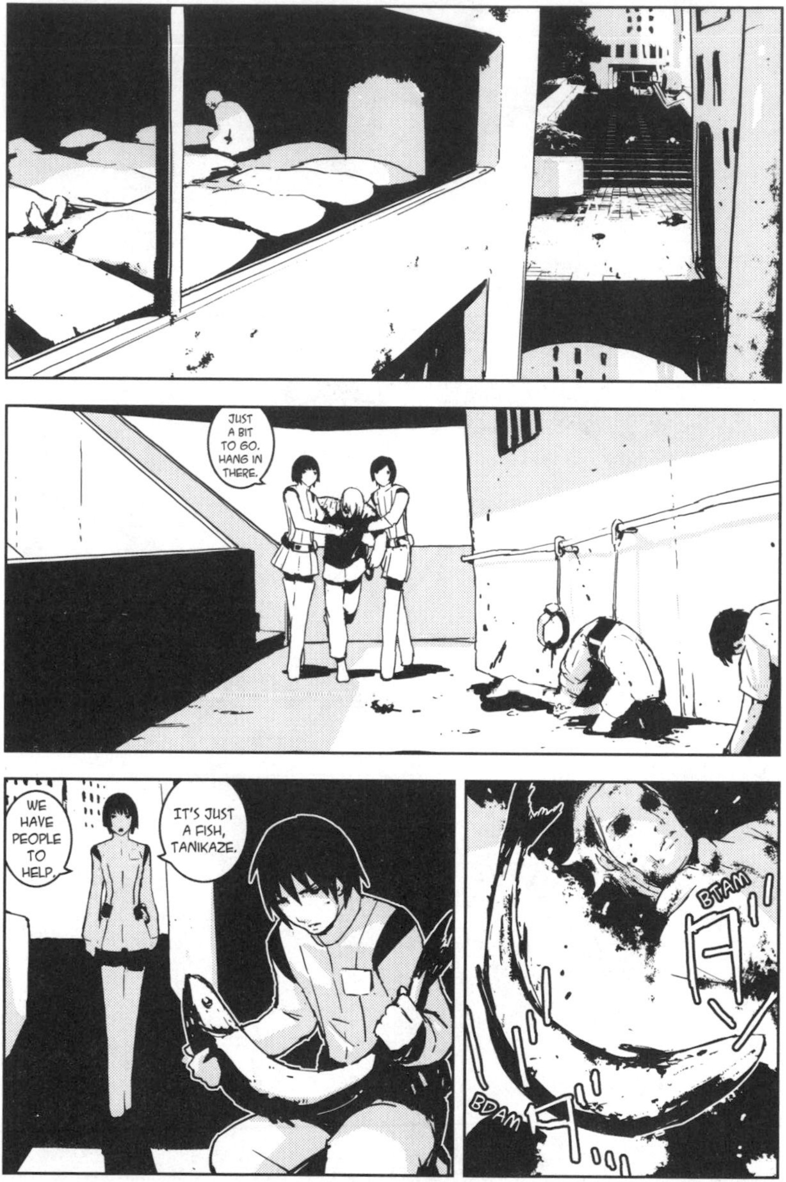
JUST A BIT TO GO. HANG IN THERE.
BTAM
BDAM
IT'S JUST A FISH, TANIKAZE.
WE HAVE PEOPLE TO HELP.

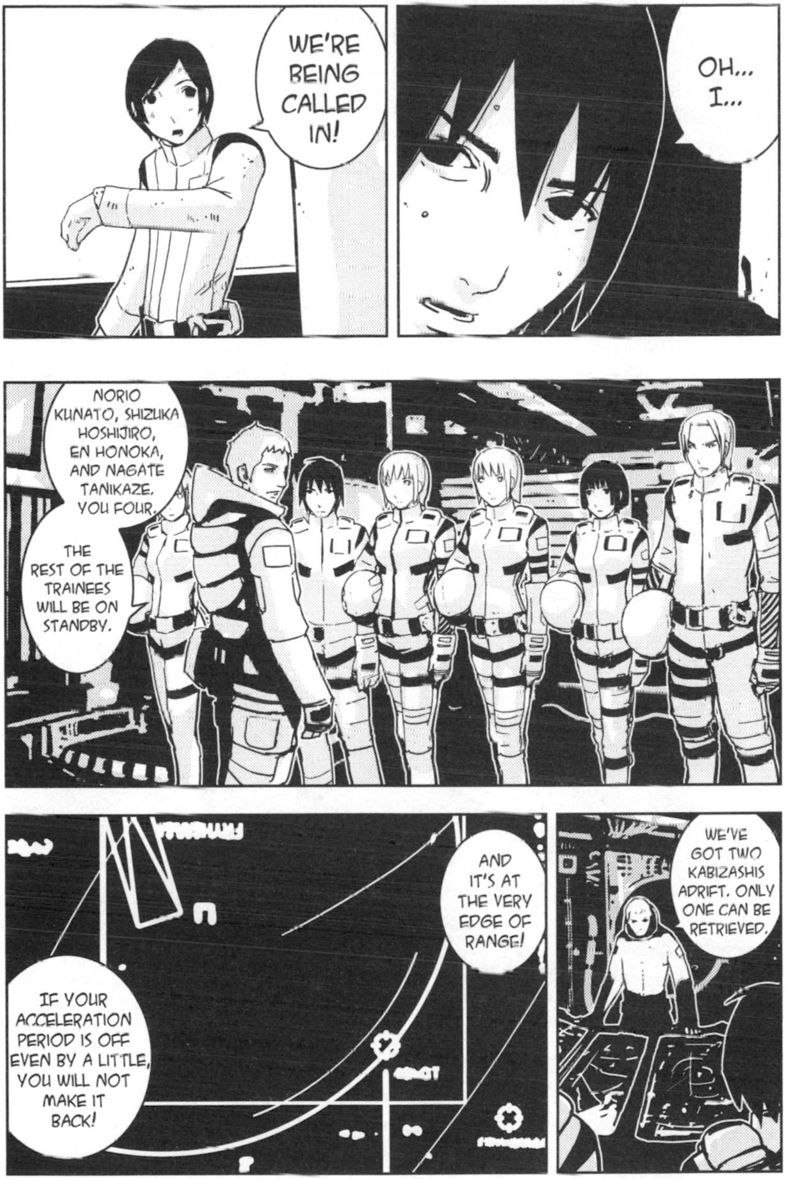
OH... I...
WE'RE BEING CALLED IN!
NORIO KUNATO, SHIZUKA HOSHIJIRO, EN HONOKA, AND NAGATE TANIKAZE. YOU FOUR.
THE REST OF THE TRAINEES WILL BE ON STANDBY.
WE'VE GOT TWO KABIZASHIS ADRIFT. ONLY ONE CAN BE RETRIEVED.
AND IT'S AT THE VERY EDGE OF RANGE!
IF YOUR ACCELERATION PERIOD IS OFF EVEN BY A LITTLE, YOU WILL NOT MAKE IT BACK!

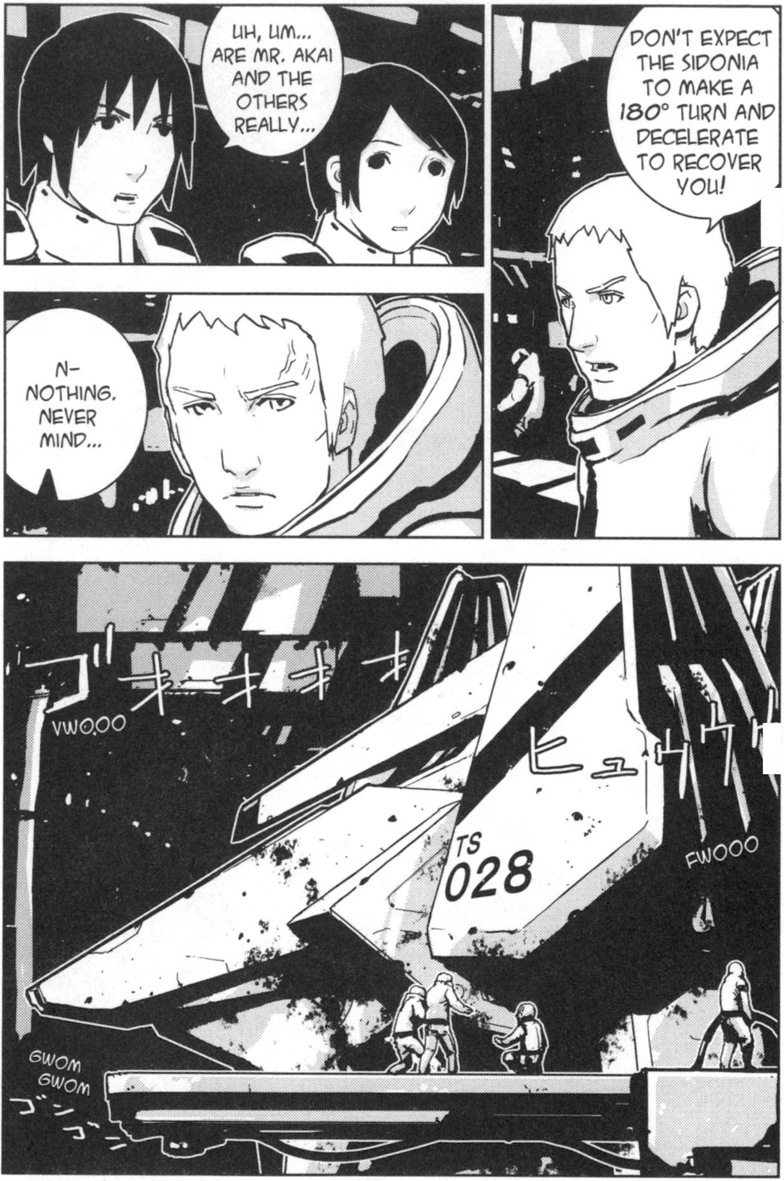
DON'T EXPECT THE SIDONIA TO MAKE A 180° TURN AND DECELERATE TO RECOVER YOU!
UH, UM... ARE MR. AKAI AND THE OTHERS REALLY...
N- NOTHING. NEVER MIND...
ヒュウウ
ゴオオオオ
VWOOO
TS 028
FWOOO
GWOM GWOM
ゴンゴン

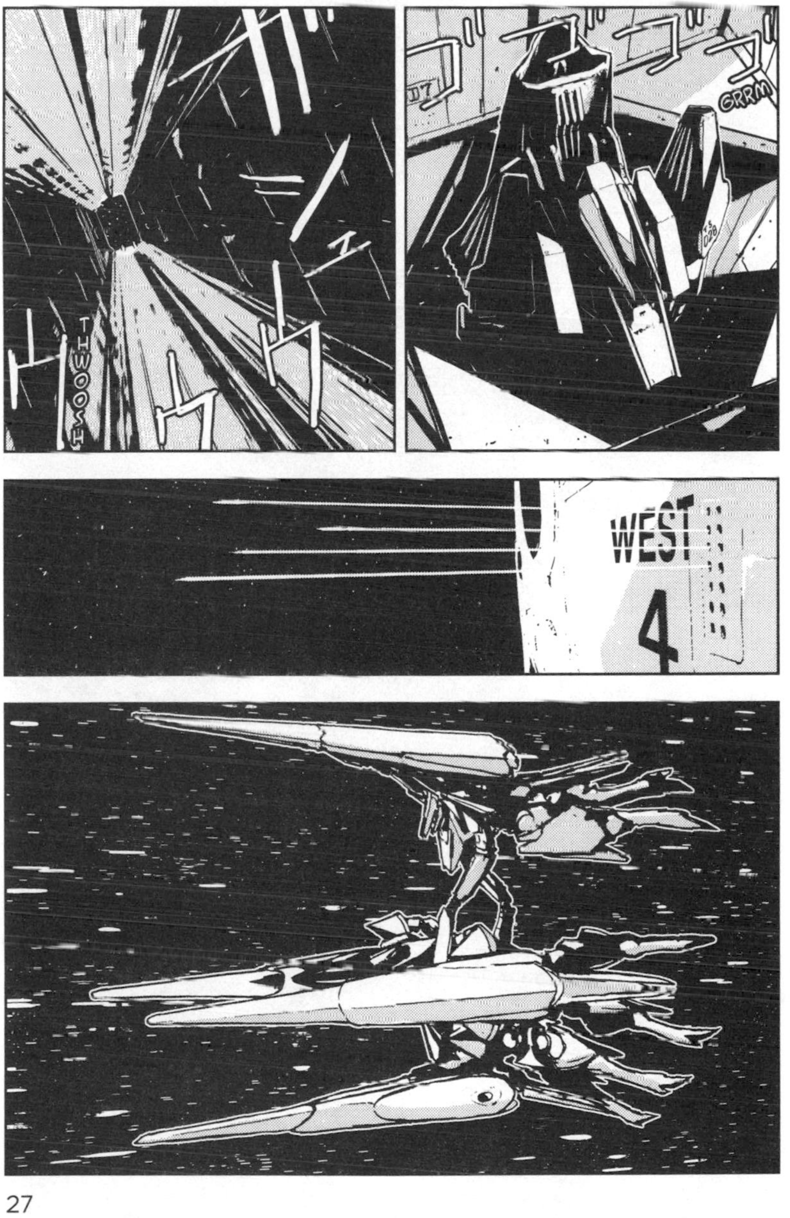
D7
GRRM
THWOOSH
WEST
4

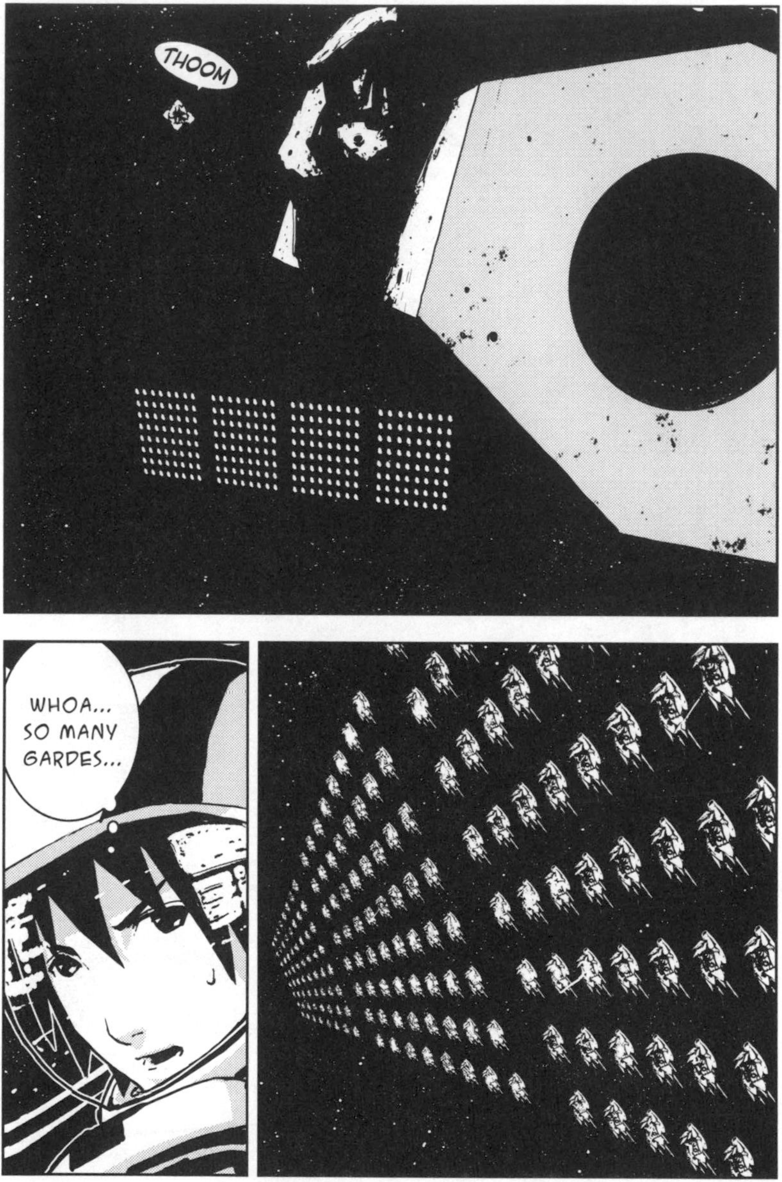
THOOM
WHOA... SO MANY GARDES...

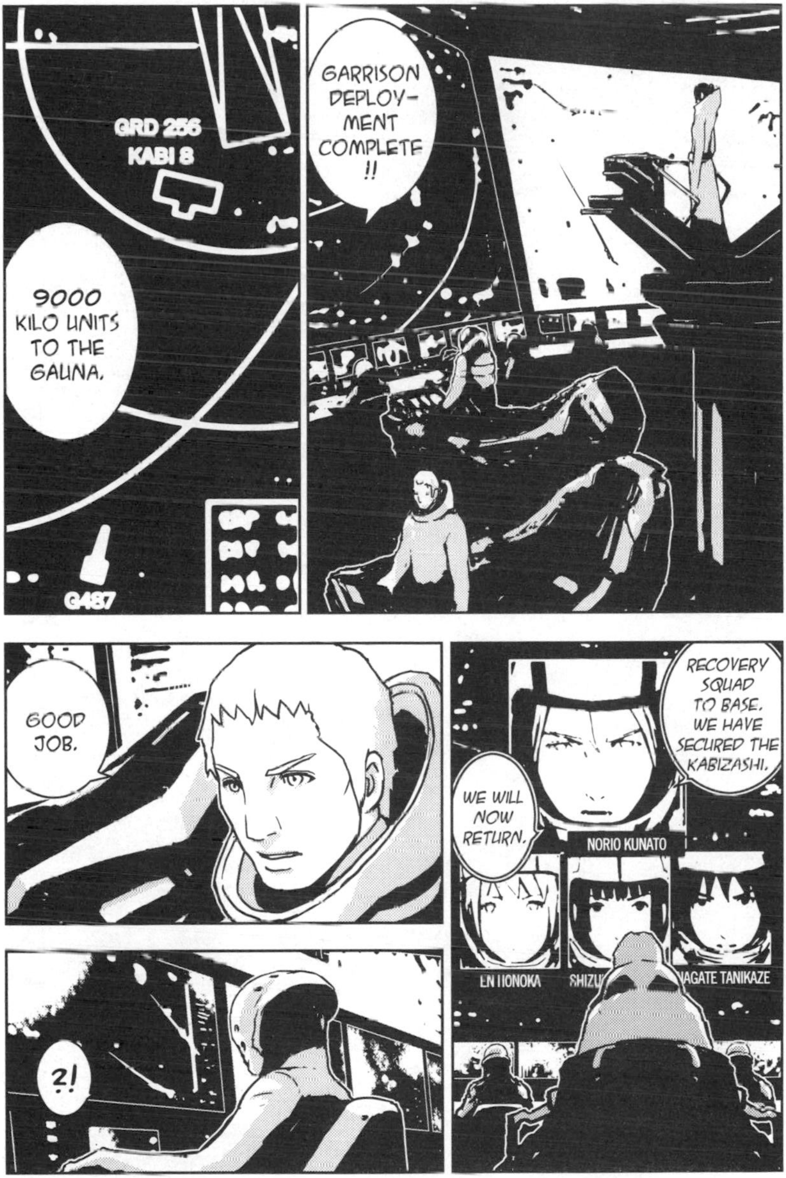
GARRISON DEPLOYMENT COMPLETE !!
GRD 256
KABI 8
9000 KILO UNITS TO THE GAUNA.
G487
RECOVERY SQUAD TO BASE. WE HAVE SECURED THE KABIZASHI.
WE WILL NOW RETURN.
NORIO KUNATO
NAGATE TANIKAZE
GOOD JOB.
?!

IT'S FACING THE RECOVERY SQUAD.

THE GALINA HAS CHANGED COURSE!!

ROGER.

ROGER.

ROGER.

THE GALINA WON'T CATCH UP WITH A FOUR-UNIT CLASP ARRAY. REMAIN CALM, AND RETURN TO BASE.

ROGER.

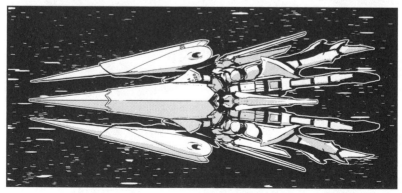

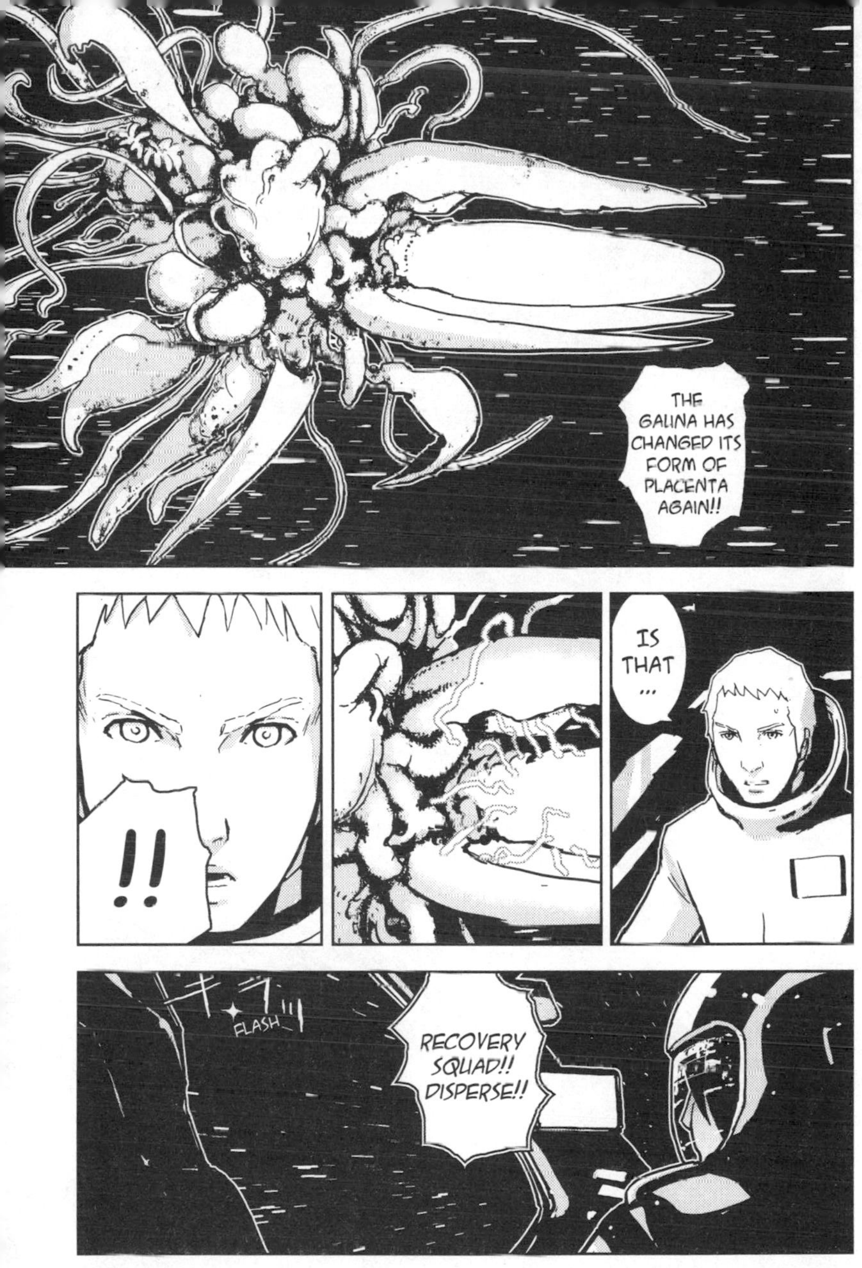
THE GALINA HAS CHANGED ITS FORM OF PLACENTA AGAIN!!
IS THAT ...
!!
RECOVERY SQUAD!! DISPERSE!!
FLASH

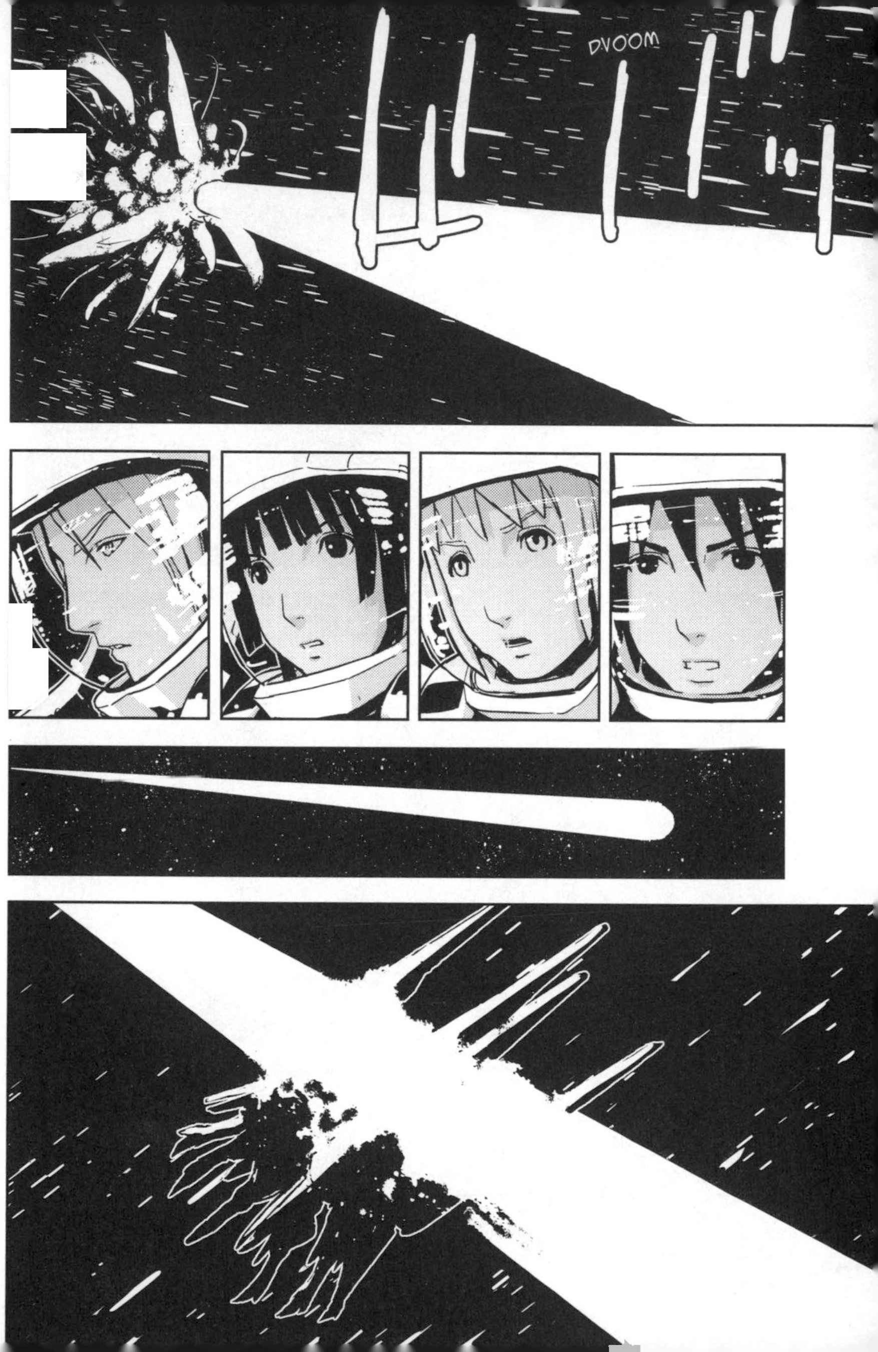
DVOOM

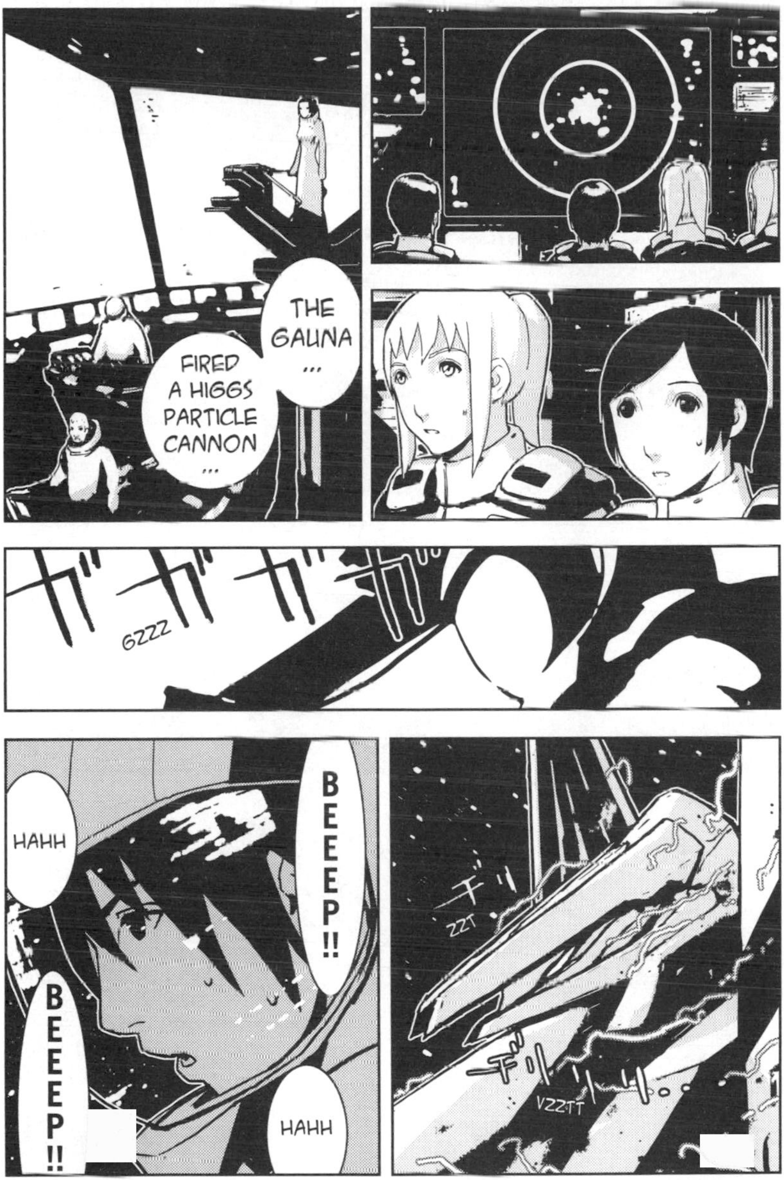
THE GAUNA ...
FIRED A HIGGS PARTICLE CANNON ...
GZZZ
ZZT
VZZTT
BEEEP!!
HAHH
BEEEP!!
HAHH

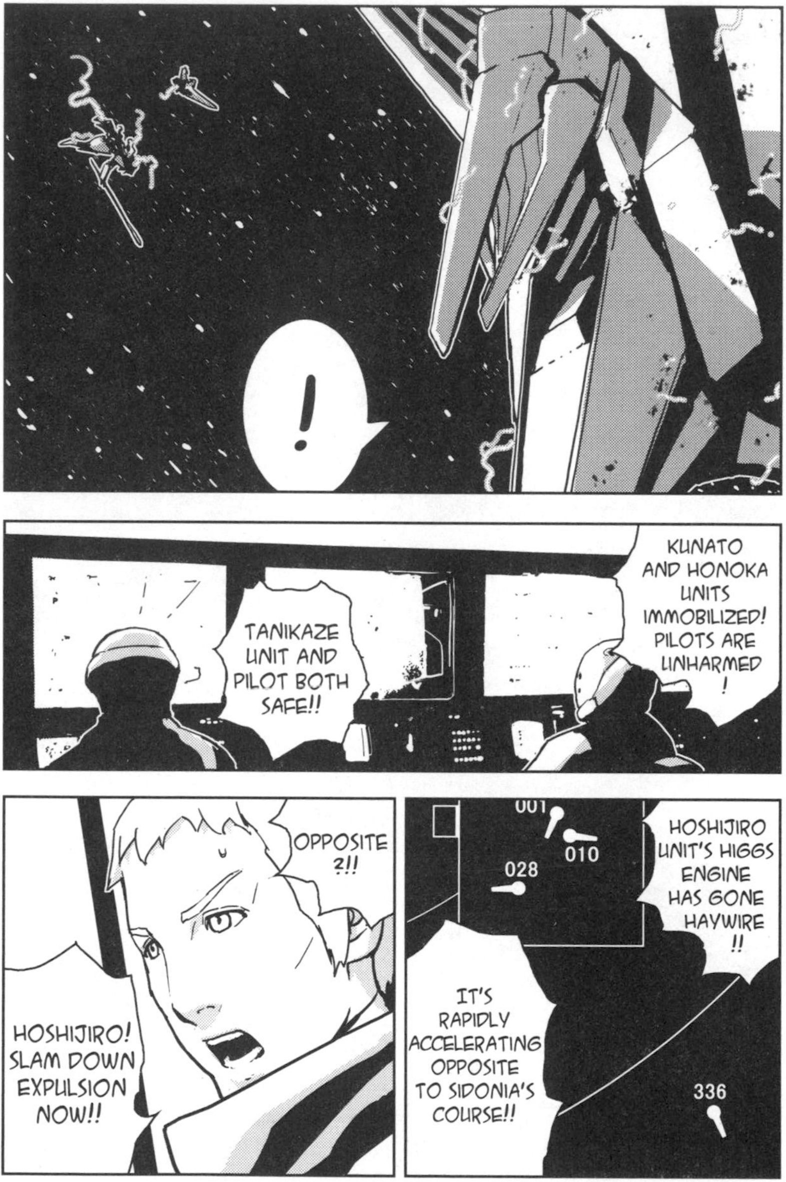
!
KUNATO AND HONOKA UNITS IMMOBILIZED! PILOTS ARE UNHARMED!
TANIKAZE UNIT AND PILOT BOTH SAFE!!
HOSHIJIRO UNIT'S HIGGS ENGINE HAS GONE HAYWIRE!!
001
010
028
IT'S RAPIDLY ACCELERATING OPPOSITE TO SIDONIA'S COURSE!!
336
OPPOSITE?!!
HOSHIJIRO! SLAM DOWN EXPULSION NOW!!

I CAN'T, CONTROLS ARE DOWN !!
VOOM
DETATCH REAR JET SYSTEM!!
IT'S NOT RESPONDING !!
RIO KUNATO
HOSHIJIRO UNIT'S HIGGS ENGINE IS AT CRITICAL LEVELS!!
SHIZUKA HOSHIJIRO
NAGATE TANIKAZE
I'M EJECT-ING!!
BOMF

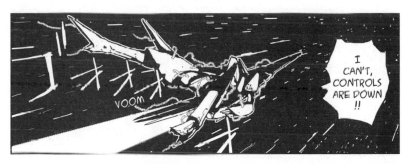

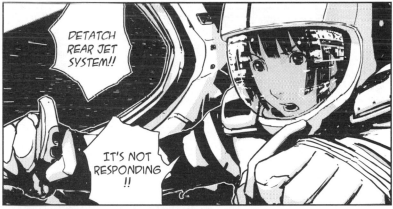

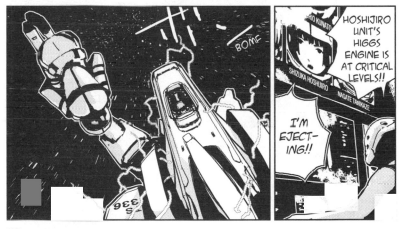

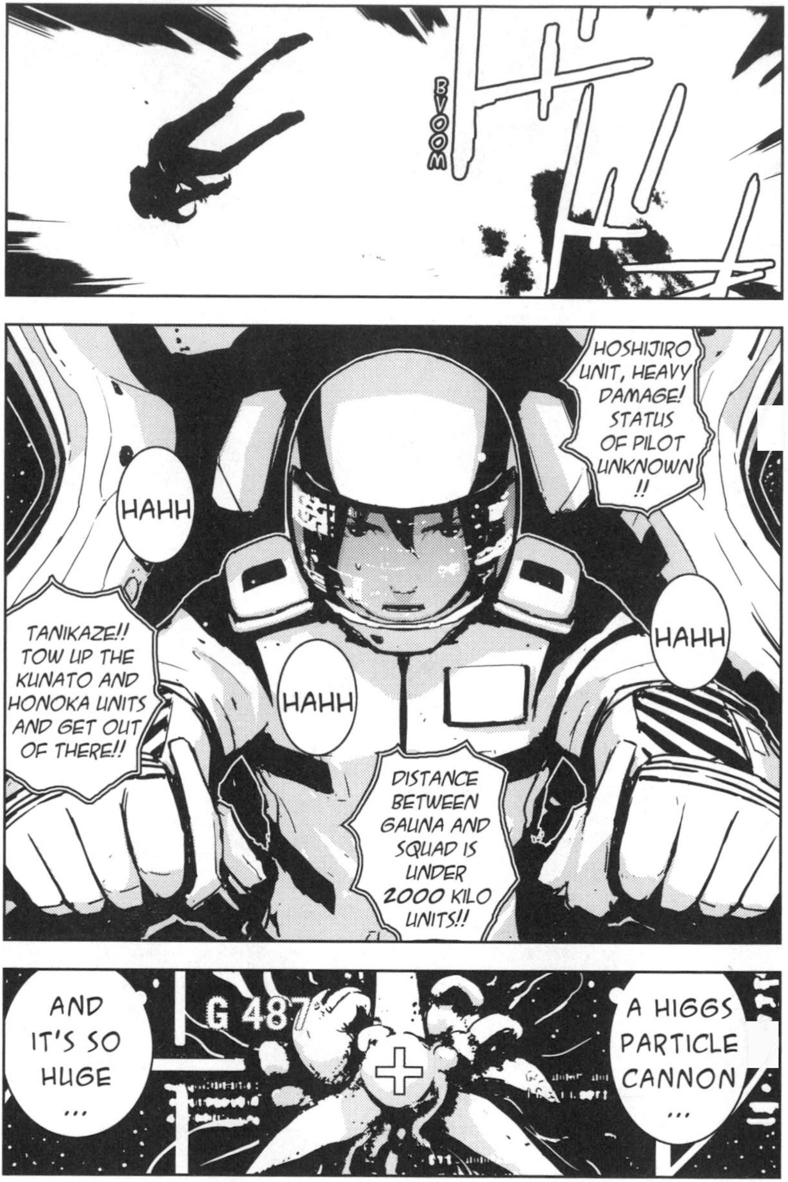
BVOOM
HOSHIJIRO UNIT, HEAVY DAMAGE! STATUS OF PILOT UNKNOWN !!
HAHH
HAHH
TANIKAZE!! TOW UP THE KUNATO AND HONOKA UNITS AND GET OUT OF THERE!!
HAHH
DISTANCE BETWEEN GAUNA AND SQUAD IS UNDER 2000 KILO UNITS!!
A HIGGS PARTICLE CANNON ...
G 487
AND IT'S SO HUGE ...

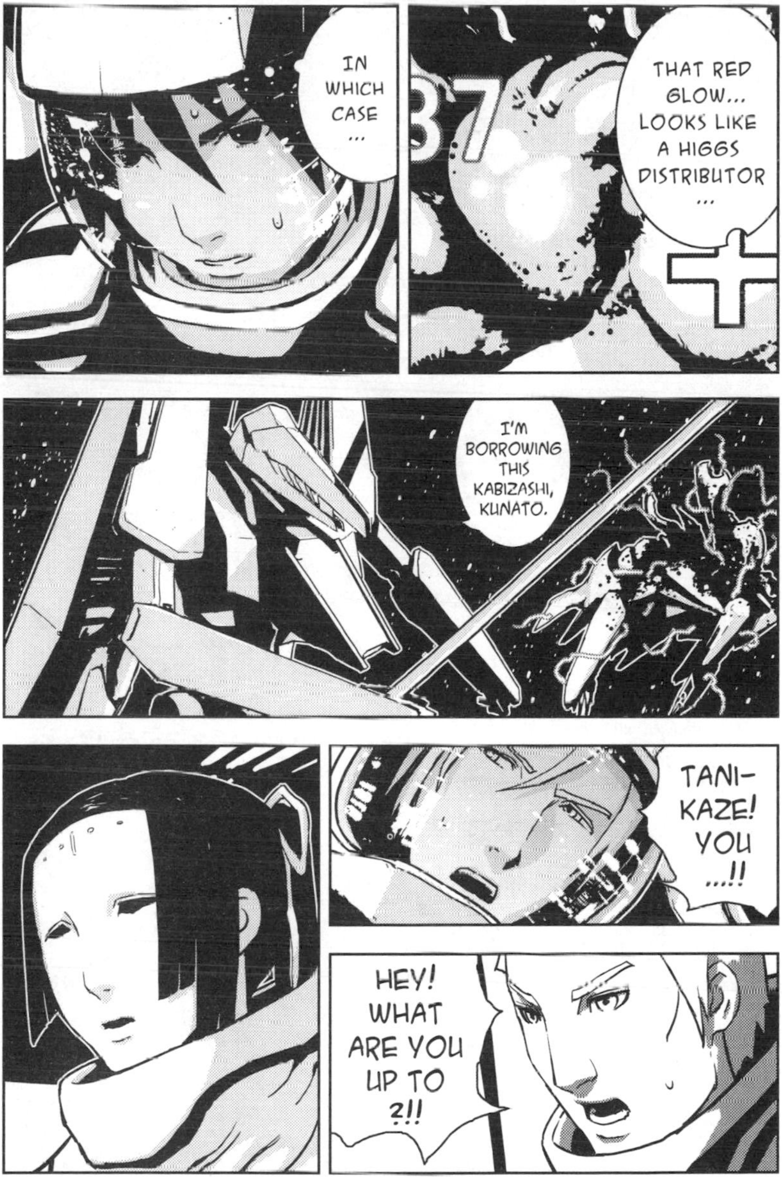
THAT RED GLOW... LOOKS LIKE A HIGGS DISTRIBUTOR ...
IN WHICH CASE ...
I'M BORROWING THIS KABIZASHI, KUNATO.
TANI-KAZE! YOU ...!!
HEY! WHAT ARE YOU UP TO ?!!

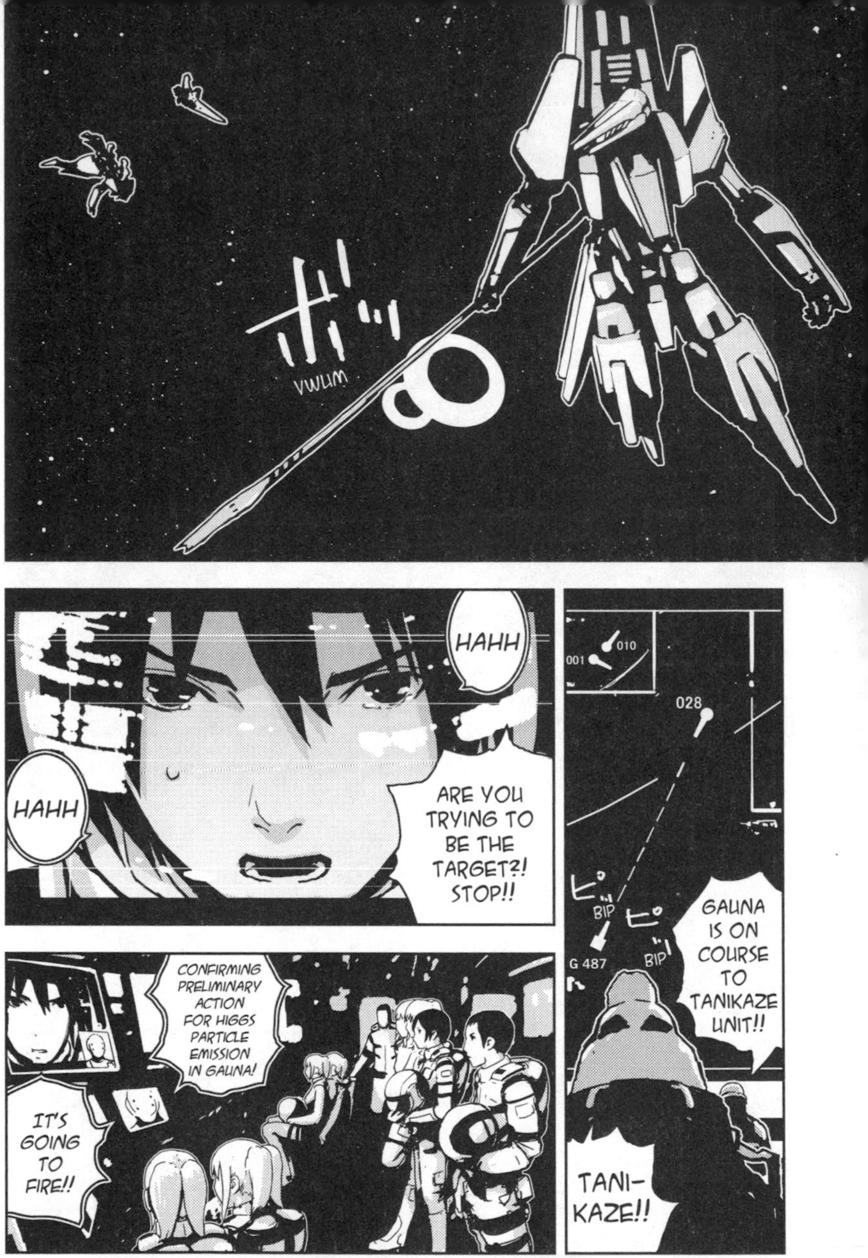
ボッ
VWUM
001
010
028
ピッ
BIP
ピッ
BIP
G 487
GAUNA IS ON COURSE TO TANIKAZE UNIT!!
TANI-KAZE!!
HAHH
ARE YOU TRYING TO BE THE TARGET?! STOP!!
HAHH
CONFIRMING PRELIMINARY ACTION FOR HIGGS PARTICLE EMISSION IN GAUNA!
IT'S GOING TO FIRE!!

HAHH
NAGATE, WHY...
HAHH
HAHH
W4 TS 028
W4 TS 028
FLASH
THE GAUNA HAS FIRED!!
028
AAH !!

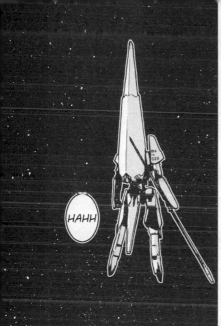

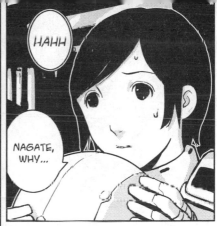

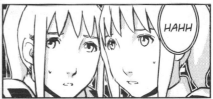

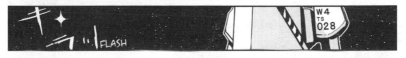

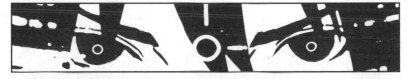

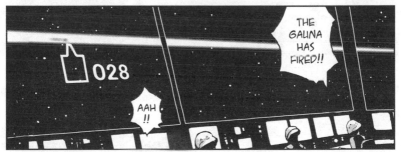

VWSH
KTANK
VZZRR
VWUM

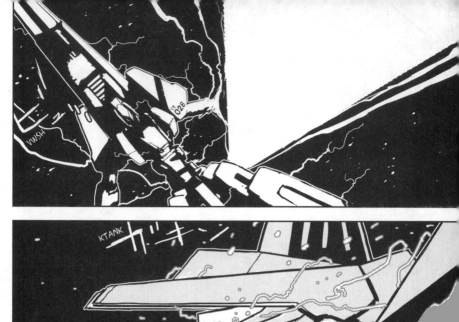

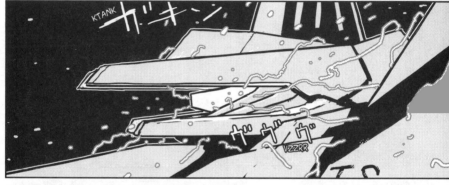

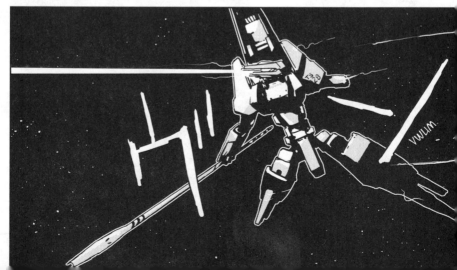

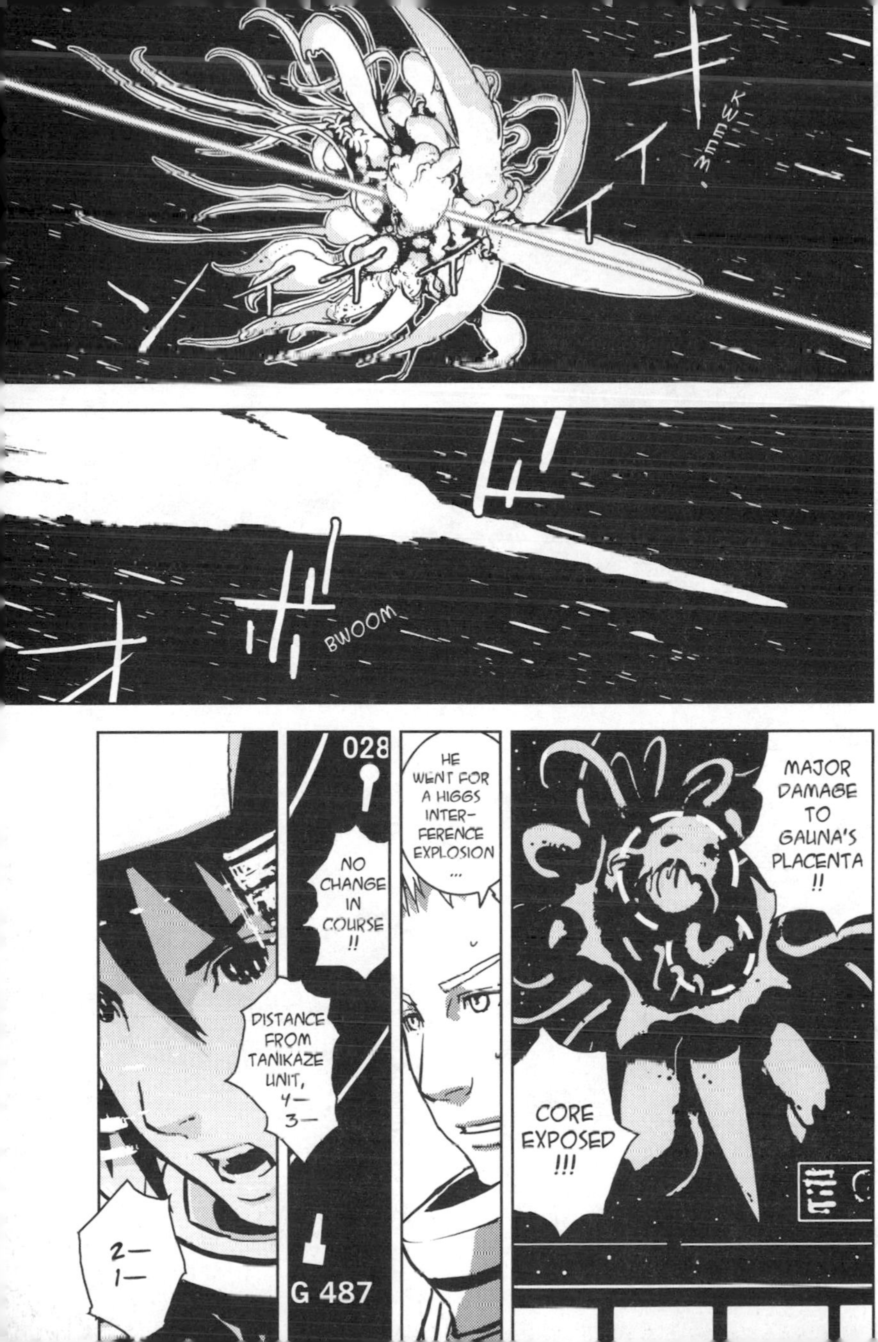
KWEEEE
BWOOM
MAJOR DAMAGE TO GAUNA'S PLACENTA !!
CORE EXPOSED !!!
HE WENT FOR A HIGGS INTER-FERENCE EXPLOSION ...
028
NO CHANGE IN COURSE !!
G 487
DISTANCE FROM TANIKAZE UNIT, 4— 3—
2— 1—

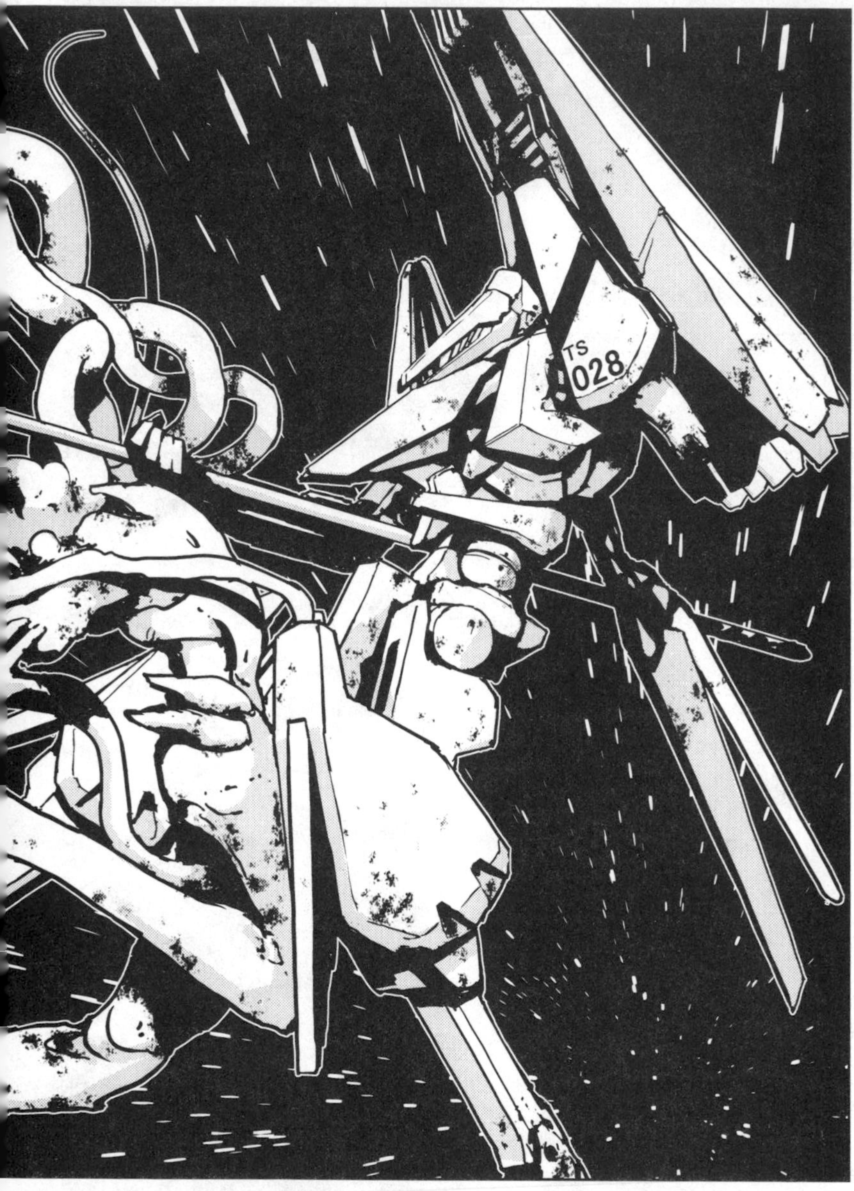
TS
028

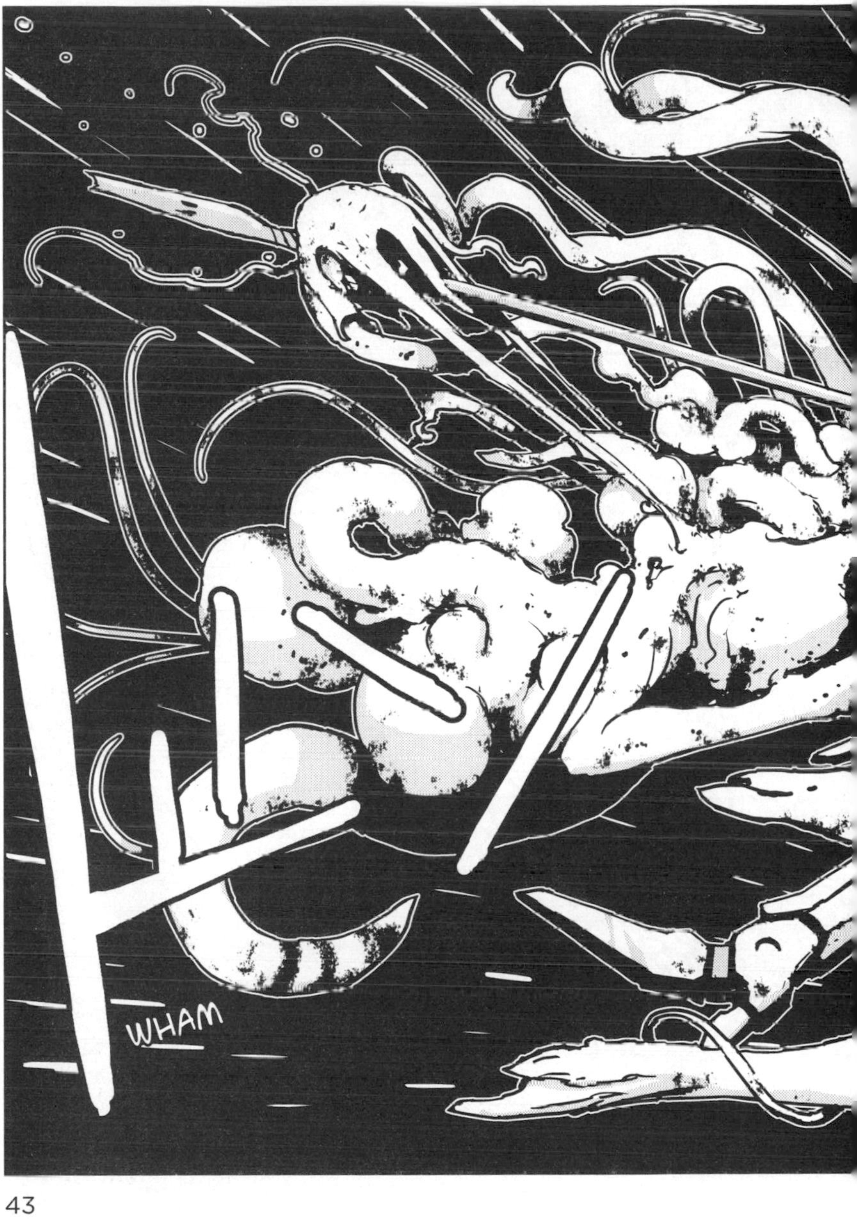
WHAM

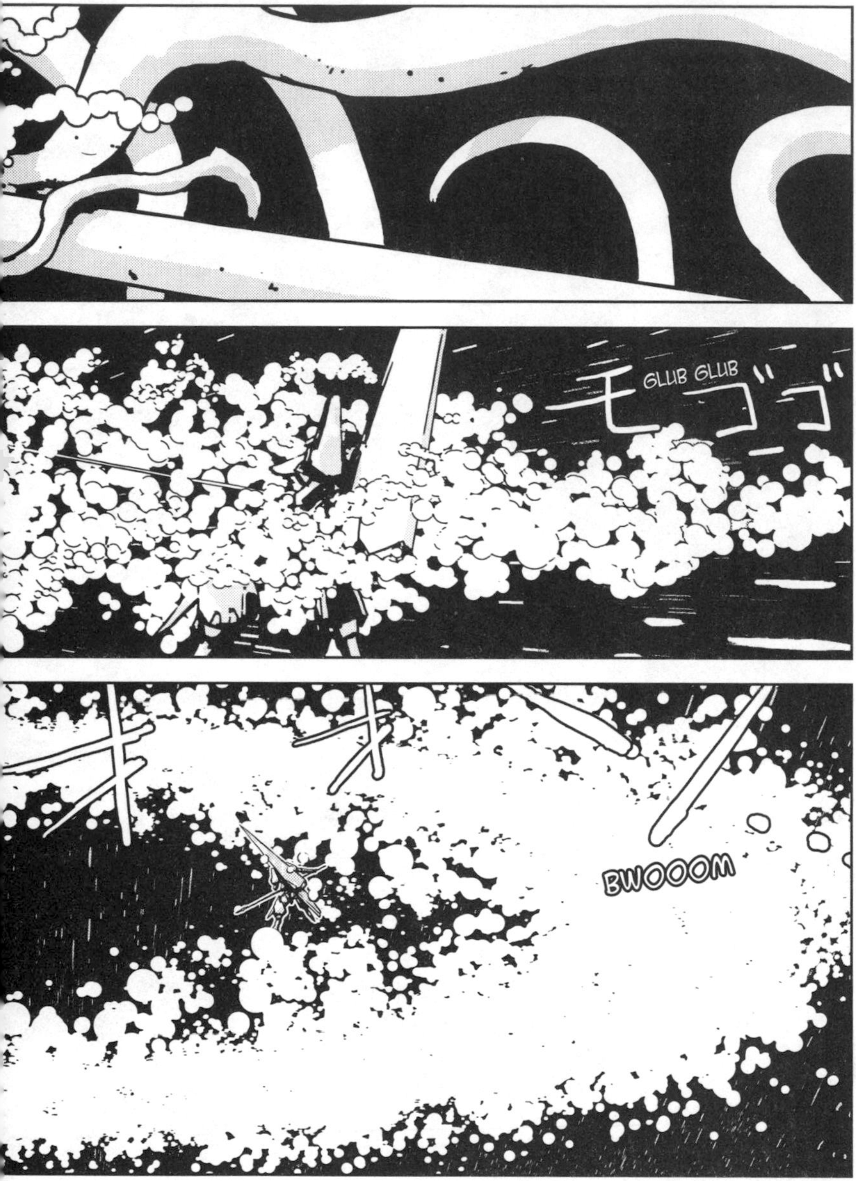
GLUB GLUB
モゴゴ
BWOOOM

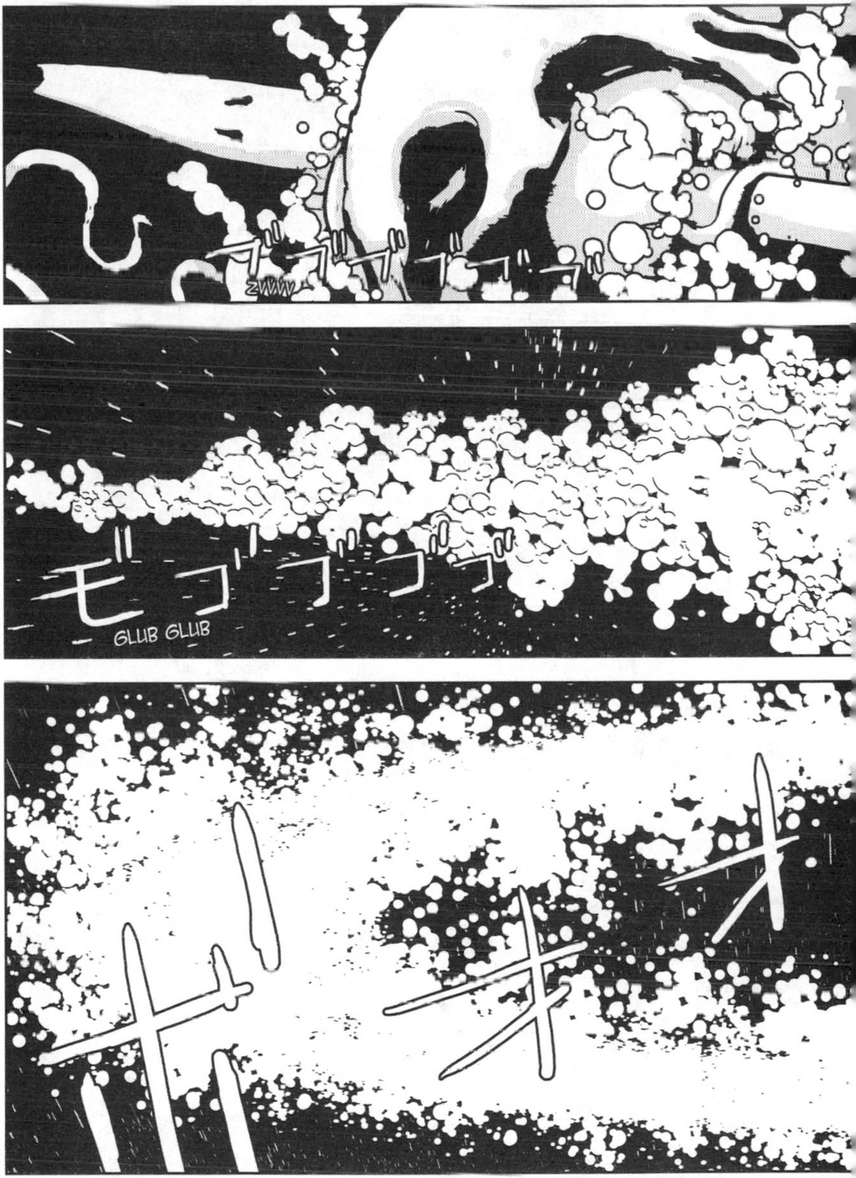
ZVVVV
GLUB GLUB

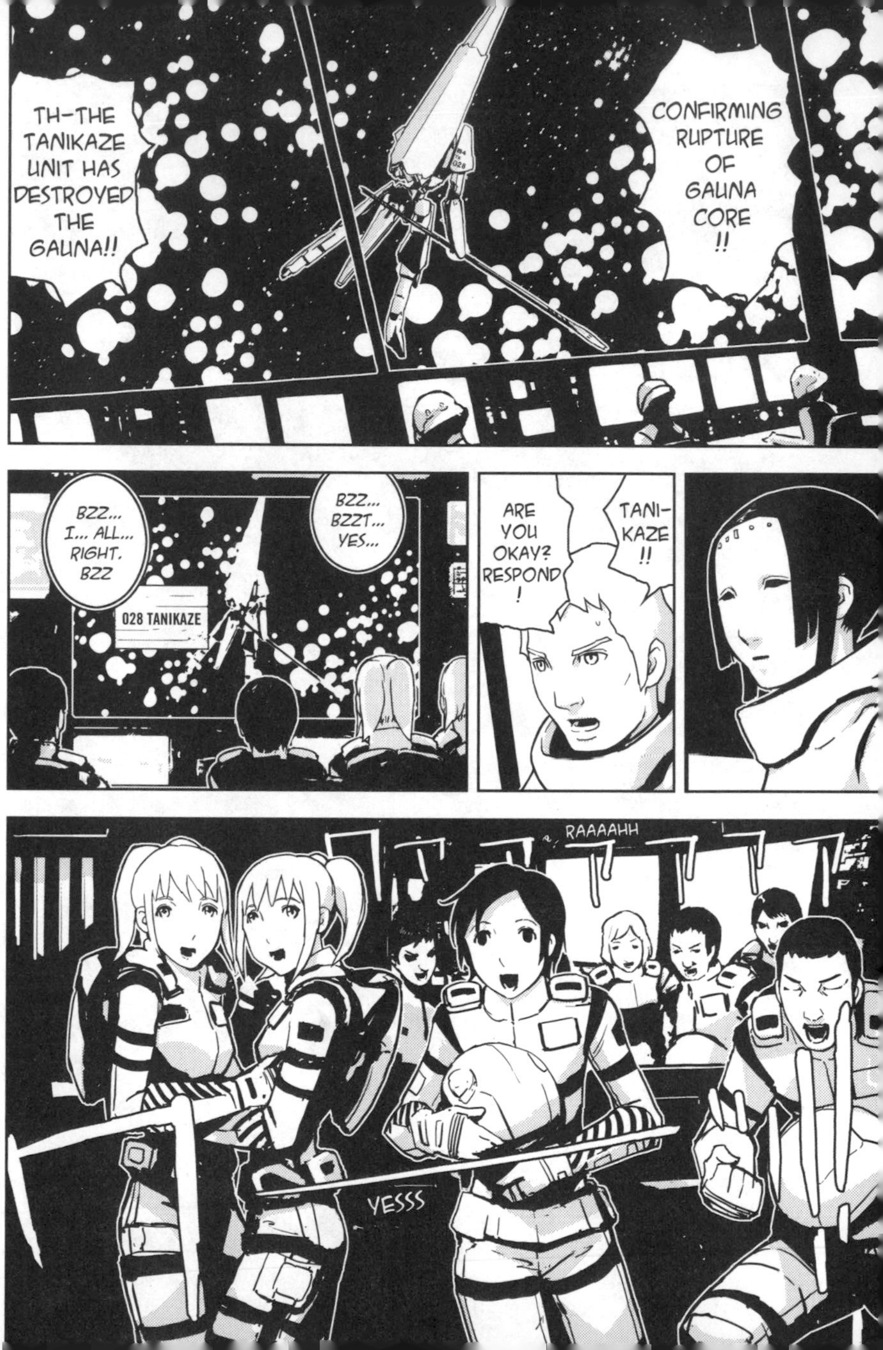
CONFIRMING RUPTURE OF GAUNA CORE!!
TH-THE TANIKAZE UNIT HAS DESTROYED THE GAUNA!!
TANI-KAZE!!
ARE YOU OKAY? RESPOND!
BZZ... BZZT... YES...
BZZ... I... ALL... RIGHT. BZZ
028 TANIKAZE
RAAAAHH
YESSS

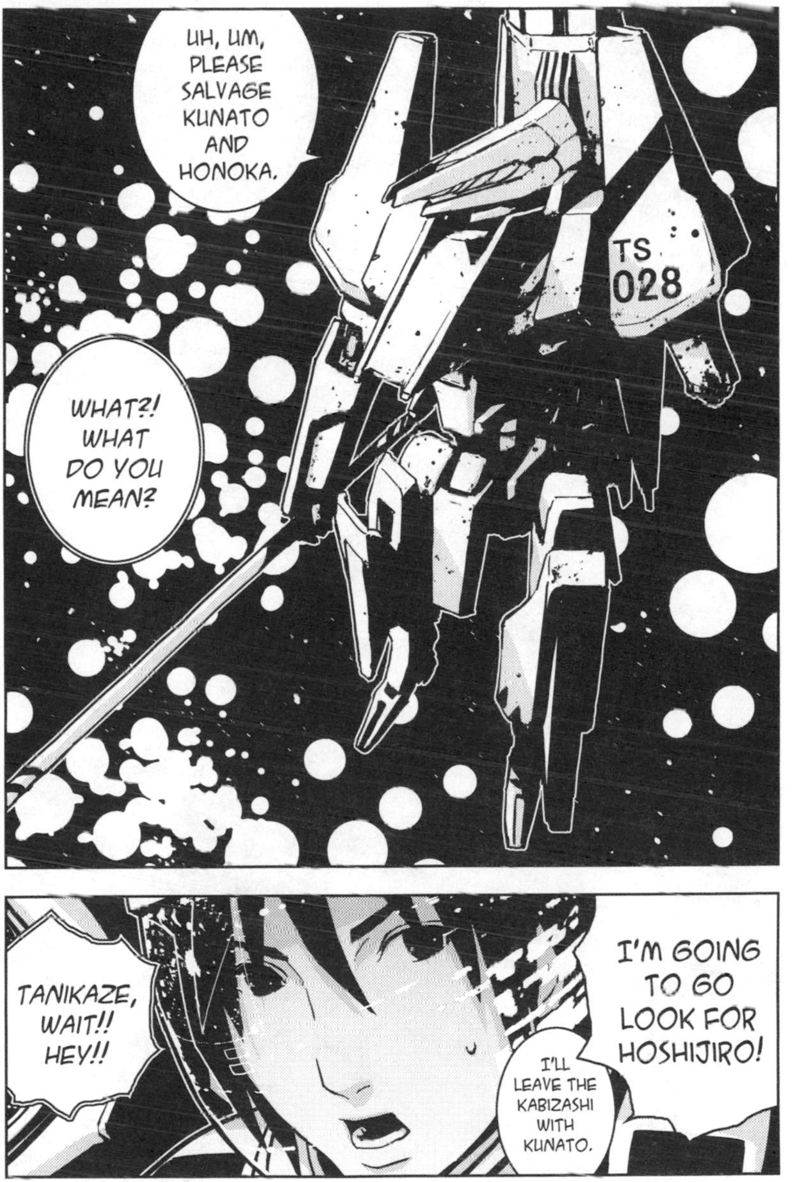
UH, UM, PLEASE SALVAGE KUNATO AND HONOKA.
TS 028
WHAT?! WHAT DO YOU MEAN?
I'M GOING TO GO LOOK FOR HOSHIJIRO!
I'LL LEAVE THE KABIZASHI WITH KUNATO.
TANIKAZE, WAIT!! HEY!!

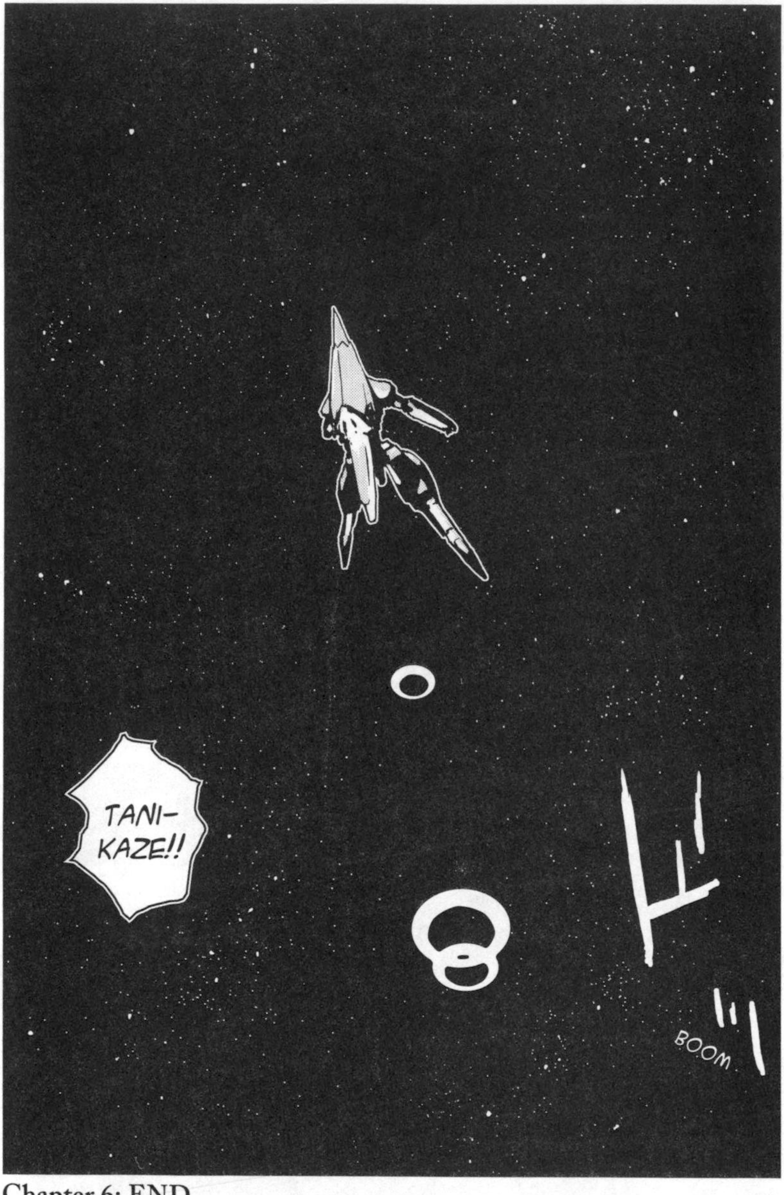
TANI-
KAZE!!
BOOM

Chapter 6: END

Chapter 7: Shizuka Hoshijiro's Rescue

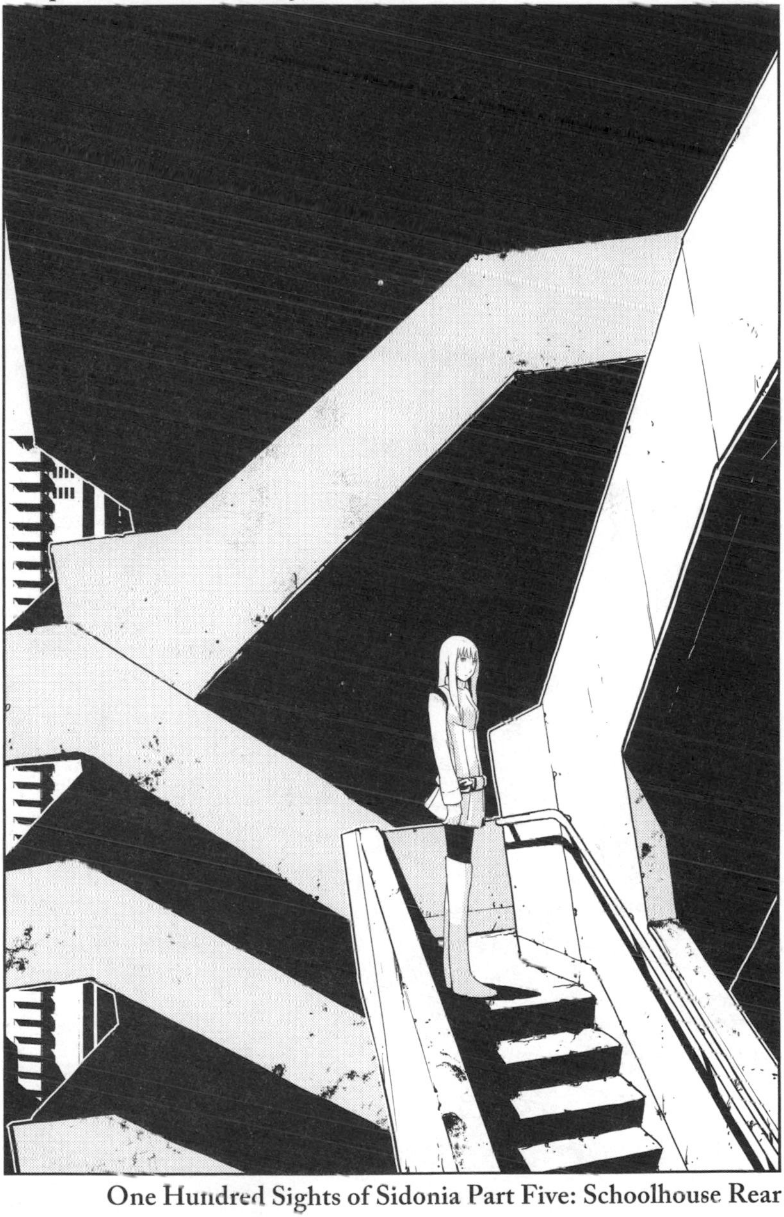

One Hundred Sights of Sidonia Part Five: Schoolhouse Rear

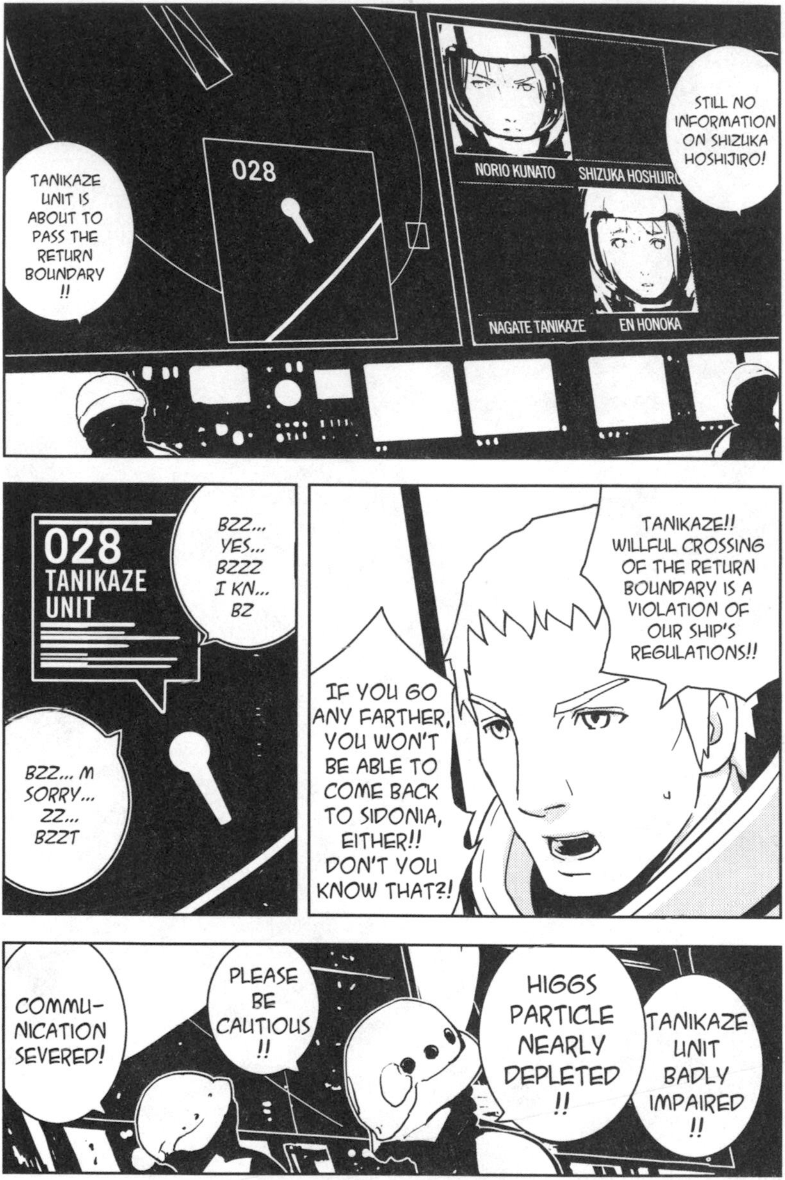
STILL NO INFORMATION ON SHIZUKA HOSHIJIRO!
NORIO KUNATO
SHIZUKA HOSHIJIRO
EN HONOKA
NAGATE TANIKAZE
028
TANIKAZE UNIT IS ABOUT TO PASS THE RETURN BOUNDARY!!
TANIKAZE!! WILLFUL CROSSING OF THE RETURN BOUNDARY IS A VIOLATION OF OUR SHIP'S REGULATIONS!!
IF YOU GO ANY FARTHER, YOU WON'T BE ABLE TO COME BACK TO SIDONIA, EITHER!! DON'T YOU KNOW THAT?!
BZZ... YES... BZZZ I KN... BZ
028
TANIKAZE UNIT
BZZ... M SORRY... ZZ... BZZT
TANIKAZE UNIT BADLY IMPAIRED!!
HIGGS PARTICLE NEARLY DEPLETED!!
PLEASE BE CAUTIOUS!!
COMMU-NICATION SEVERED!

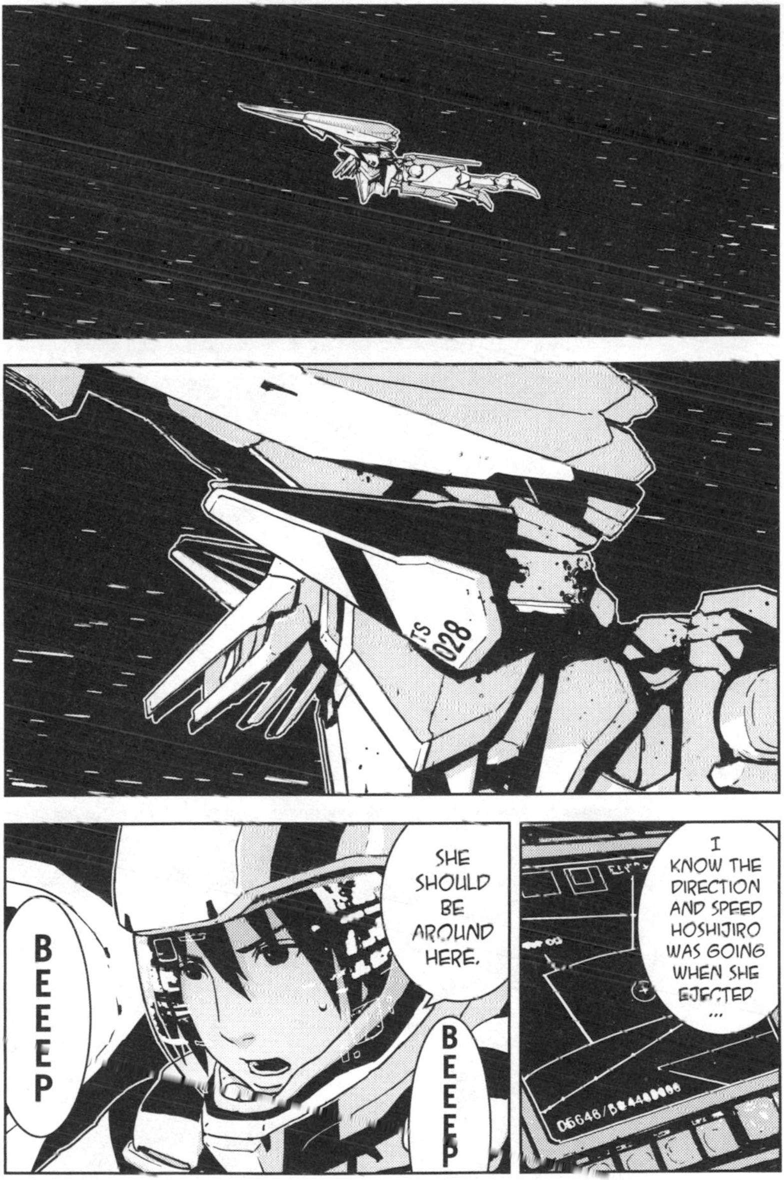
TS 028
I KNOW THE DIRECTION AND SPEED HOSHIJIRO WAS GOING WHEN SHE EJECTED ...
SHE SHOULD BE AROUND HERE.
BEEEP
BEEEP

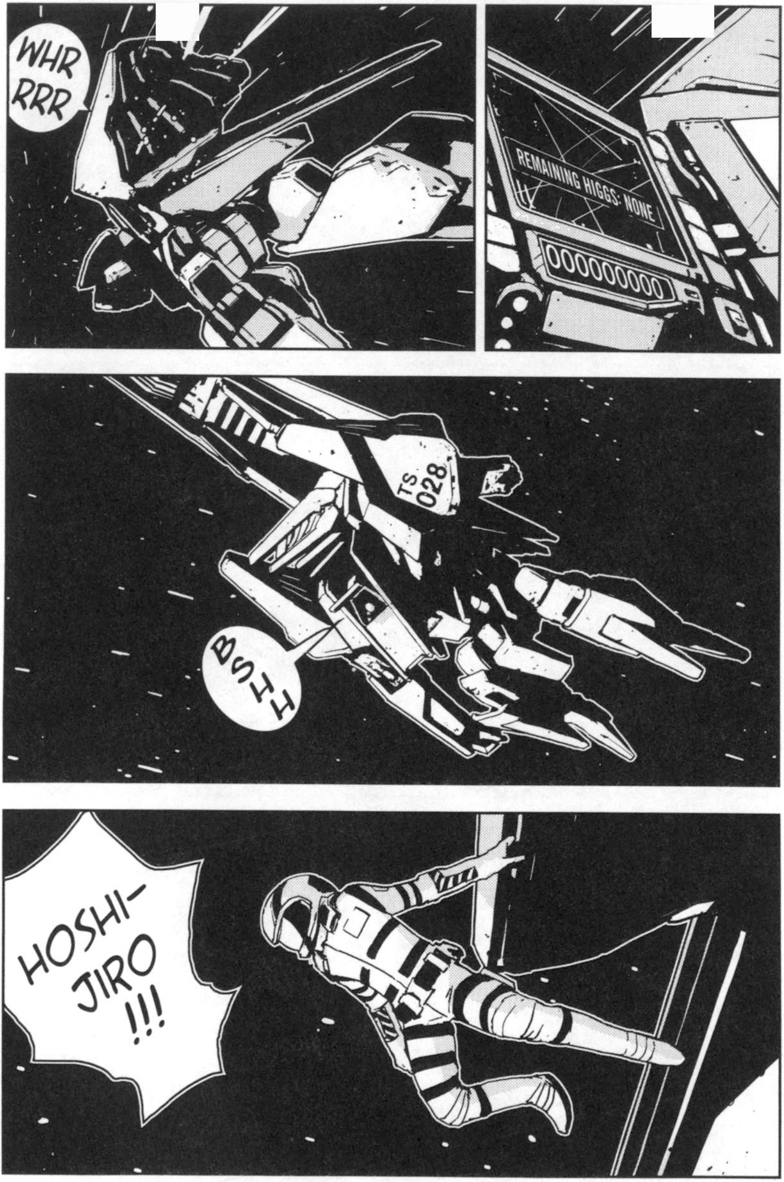
REMAINING HIGGS: NONE
WHR RRR
BSHH
HOSHI-JIRO !!!

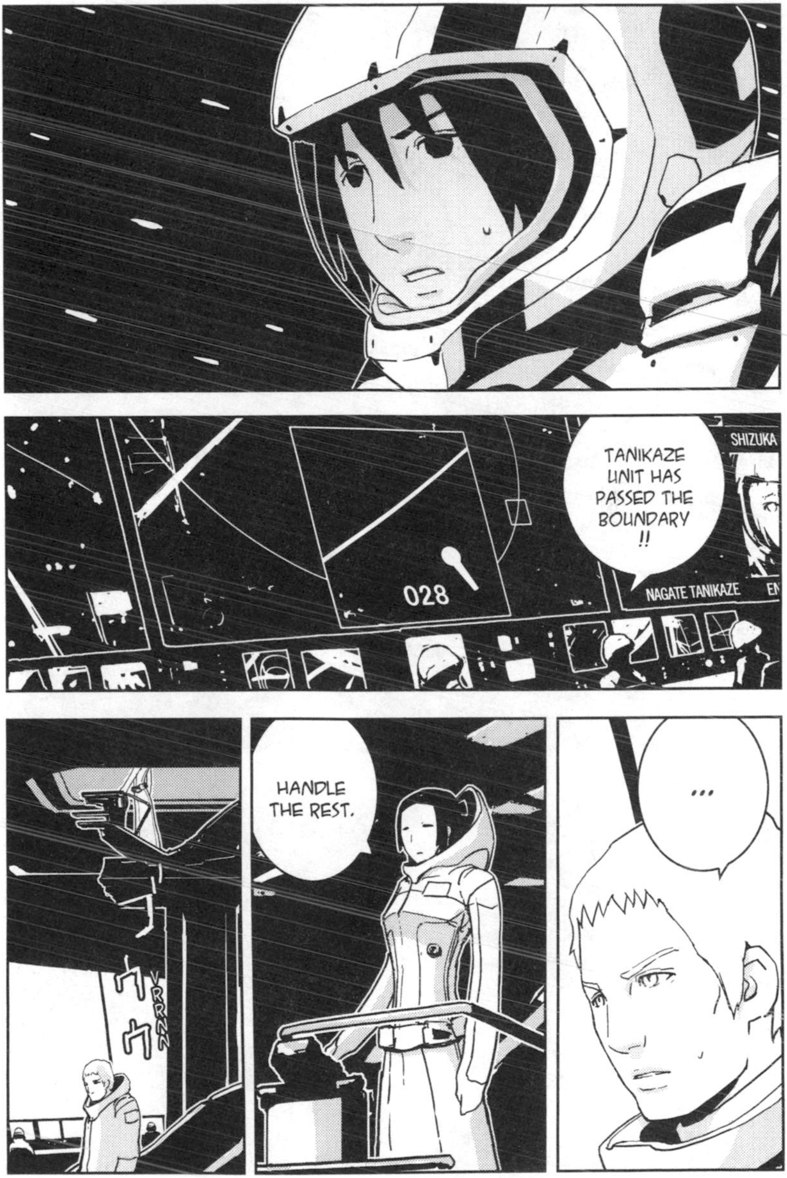
SHIZUKA
TANIKAZE UNIT HAS PASSED THE BOUNDARY !!
028
NAGATE TANIKAZE
...
HANDLE THE REST.
VRRNN

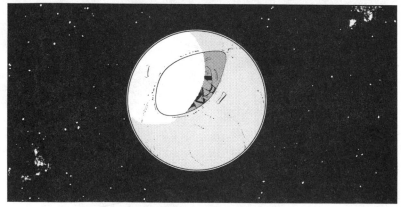

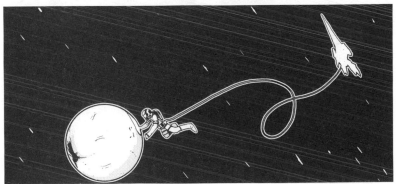

HOSHI-
JIRO
!!

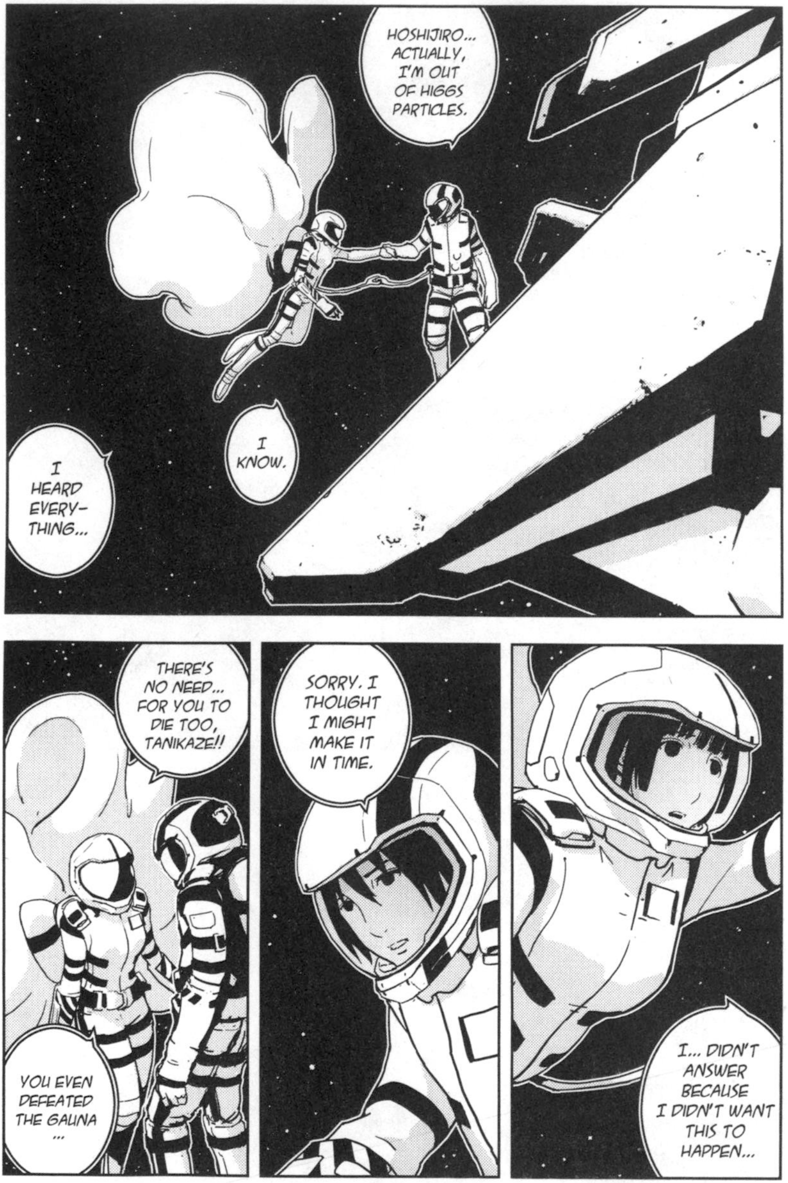
HOSHIJIRO... ACTUALLY, I'M OUT OF HIGGS PARTICLES.
I KNOW.
I HEARD EVERY-THING...
I... DIDN'T ANSWER BECAUSE I DIDN'T WANT THIS TO HAPPEN...
SORRY. I THOUGHT I MIGHT MAKE IT IN TIME.
THERE'S NO NEED... FOR YOU TO DIE TOO, TANIKAZE!!
YOU EVEN DEFEATED THE GAUNA ...

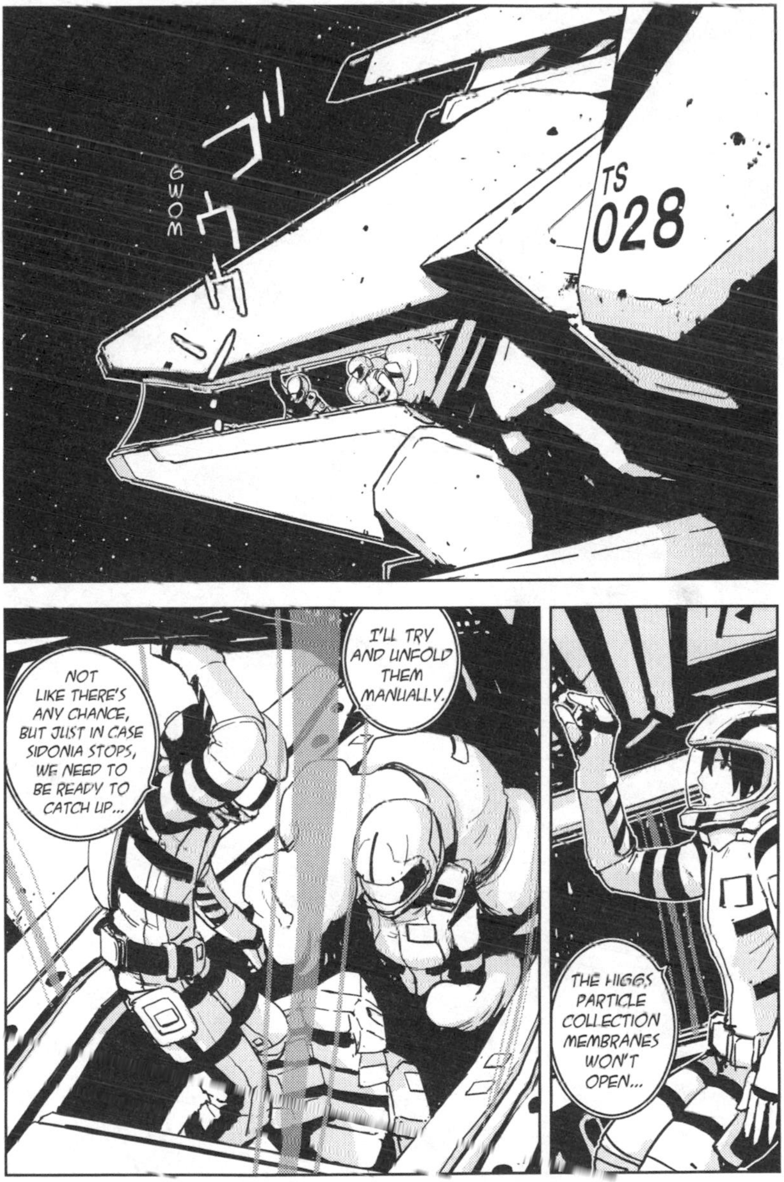
GWOM
TS 028
THE HIGGS PARTICLE COLLECTION MEMBRANES WON'T OPEN...
I'LL TRY AND UNFOLD THEM MANUALLY.
NOT LIKE THERE'S ANY CHANCE, BUT JUST IN CASE SIDONIA STOPS, WE NEED TO BE READY TO CATCH UP...

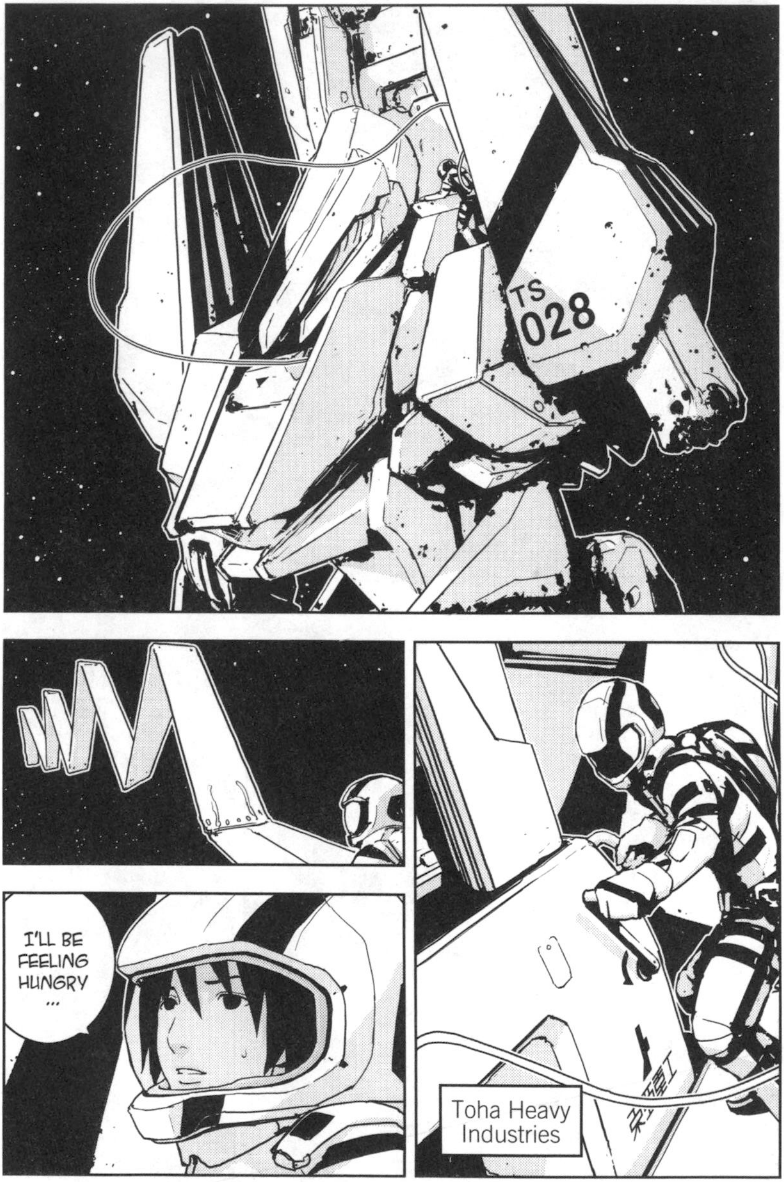
TS
028
Toha Heavy Industries
I'LL BE FEELING HUNGRY ...

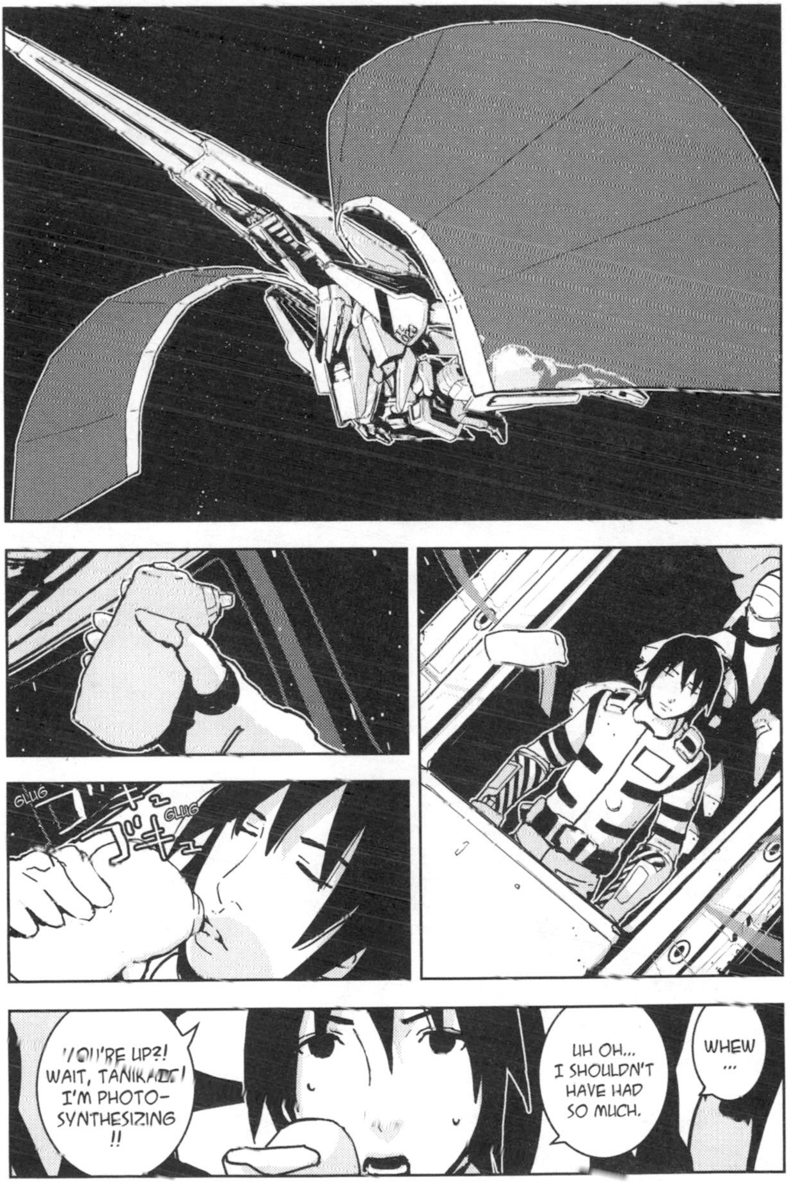
GLUG
GLUG
WHEW ...
UH OH... I SHOULDN'T HAVE HAD SO MUCH.
YOU'RE UP?! WAIT, TANIKAZE! I'M PHOTO-SYNTHESIZING !!

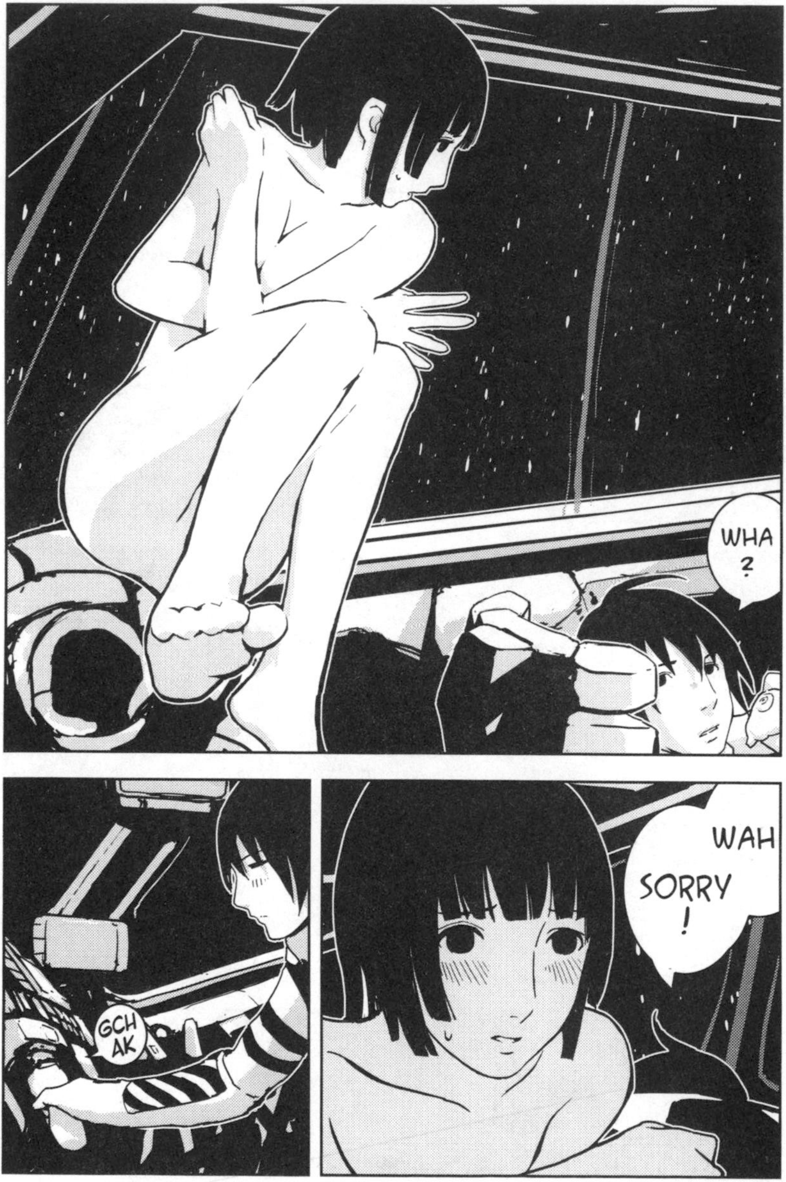
WHA ?
WAH
SORRY !
GCH AK

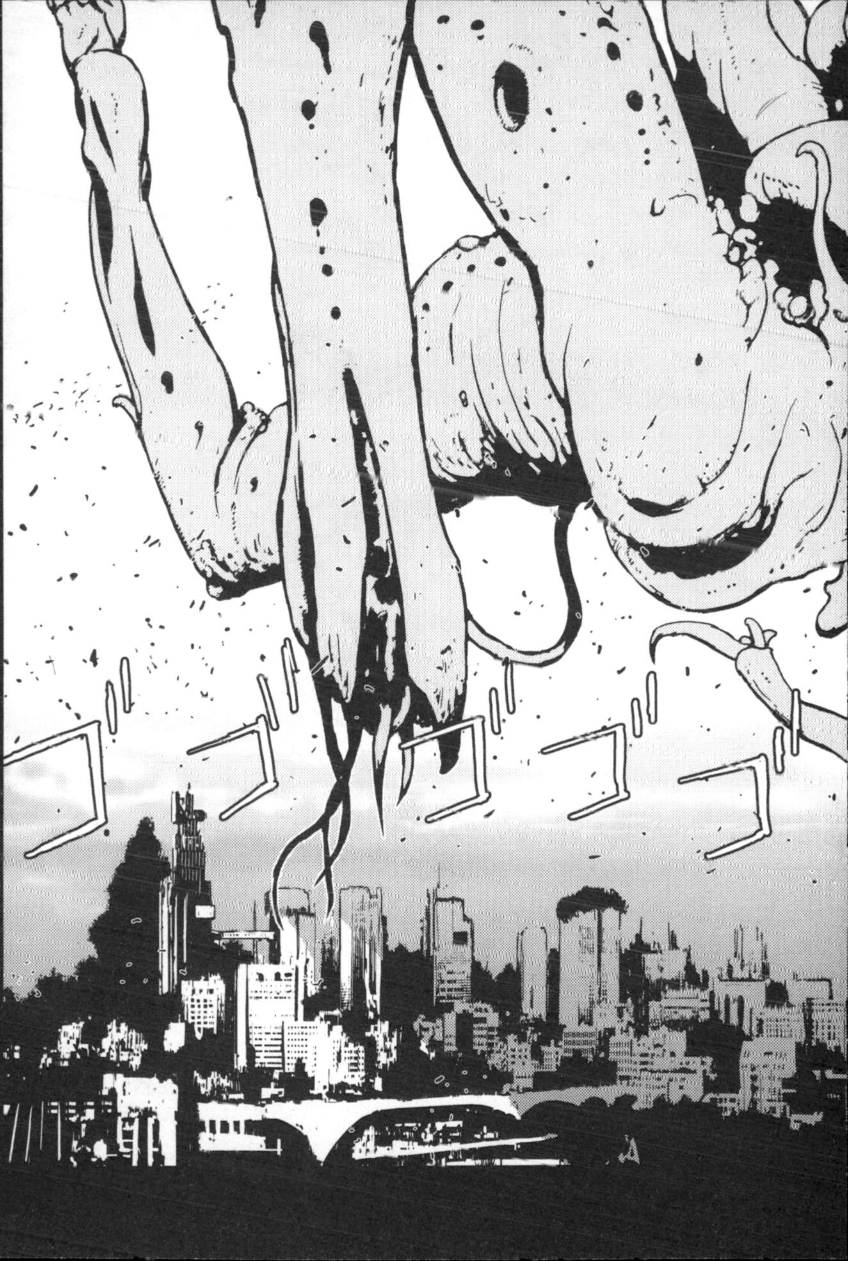
ゴゴゴゴゴ

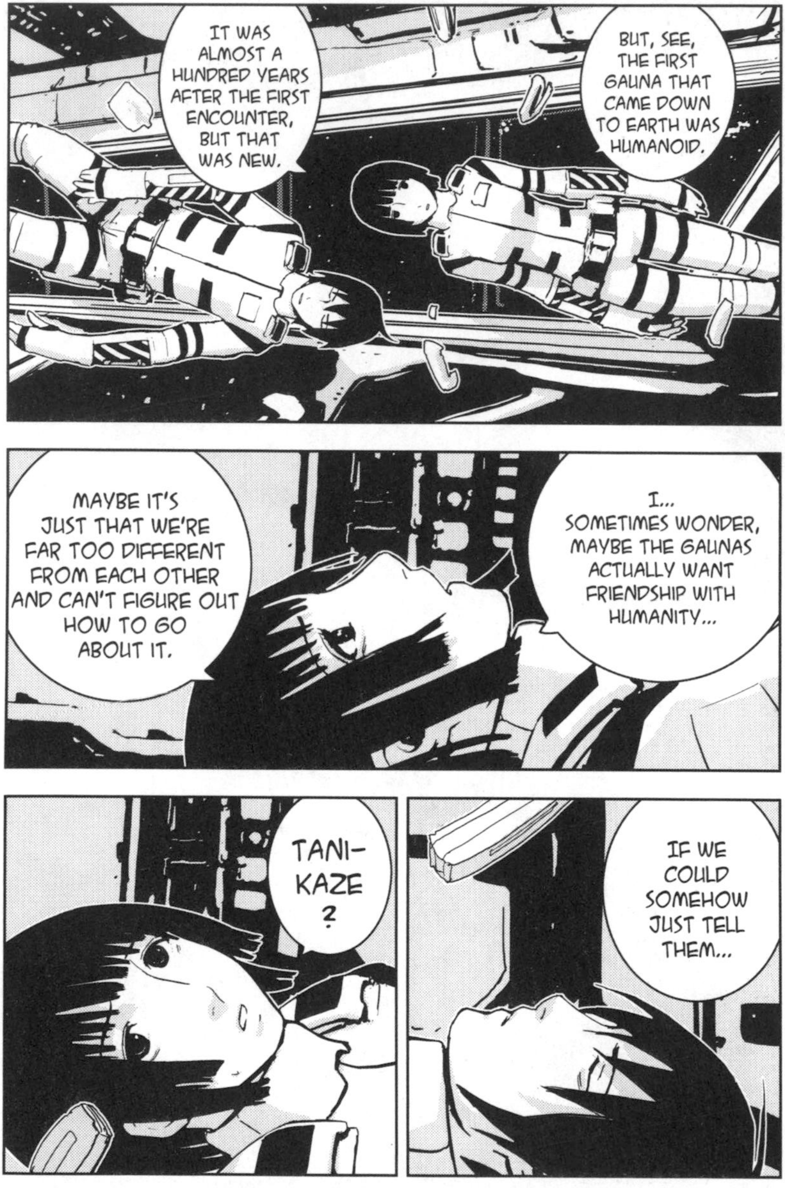
BUT, SEE, THE FIRST GAUNA THAT CAME DOWN TO EARTH WAS HUMANOID.
IT WAS ALMOST A HUNDRED YEARS AFTER THE FIRST ENCOUNTER, BUT THAT WAS NEW.
I... SOMETIMES WONDER, MAYBE THE GAUNAS ACTUALLY WANT FRIENDSHIP WITH HUMANITY...
MAYBE IT'S JUST THAT WE'RE FAR TOO DIFFERENT FROM EACH OTHER AND CAN'T FIGURE OUT HOW TO GO ABOUT IT.
IF WE COULD SOMEHOW JUST TELL THEM...
TANI-KAZE?

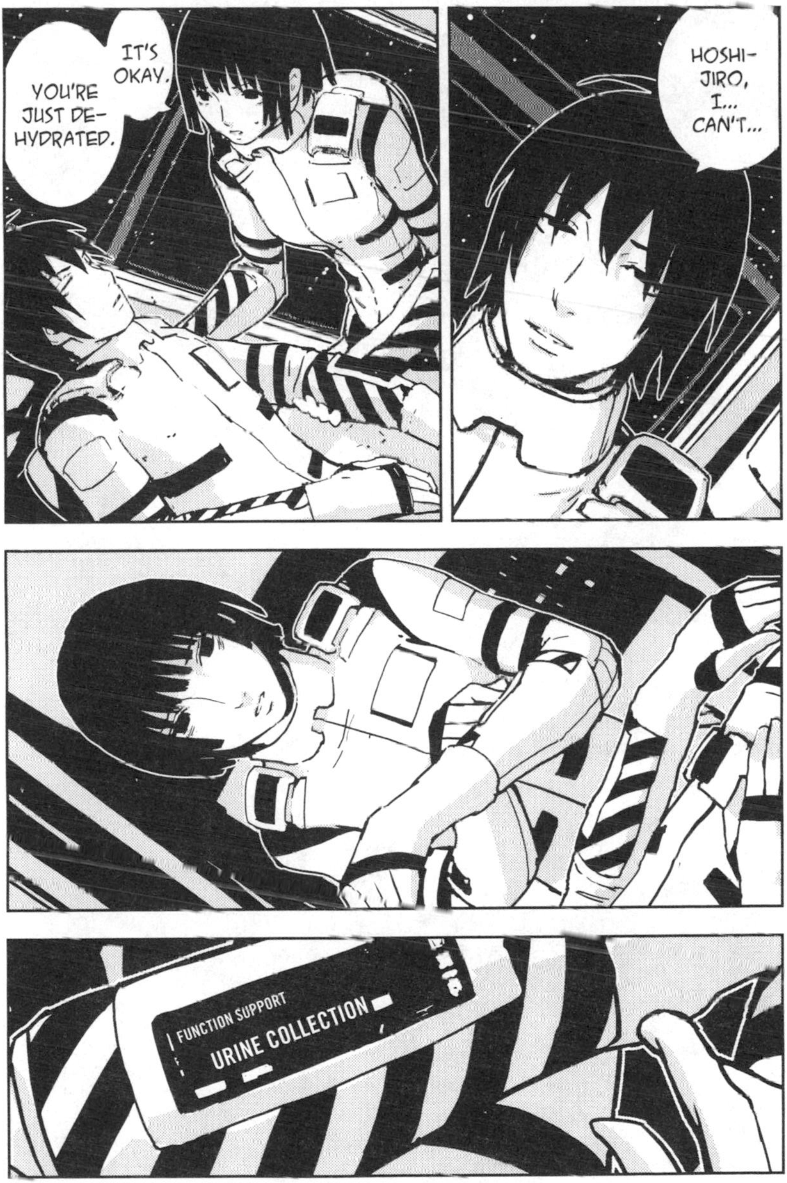
HOSHI-JIRO, I... CAN'T...
IT'S OKAY.
YOU'RE JUST DE-HYDRATED.
FUNCTION SUPPORT
URINE COLLECTION

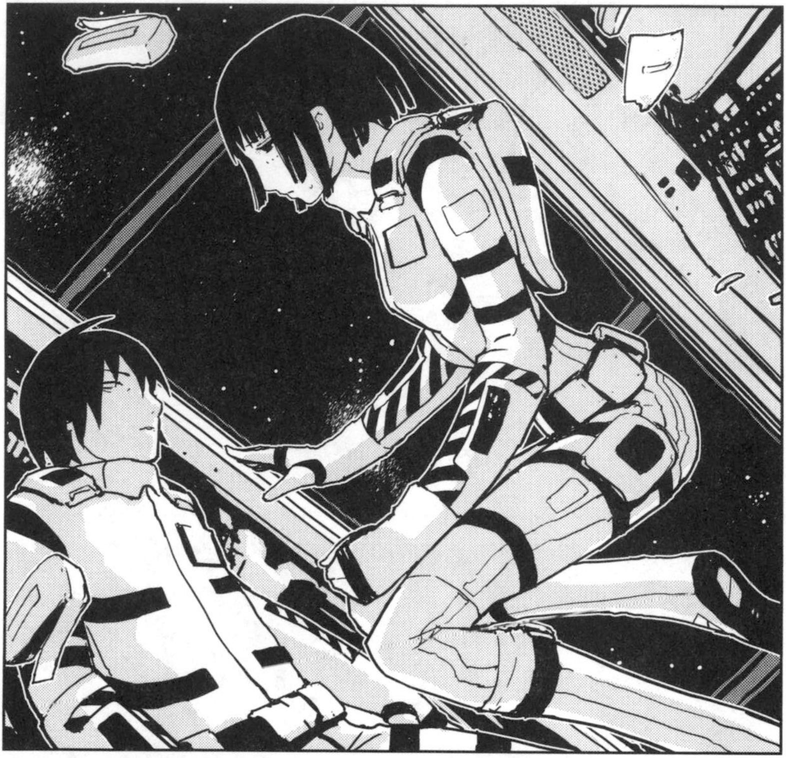

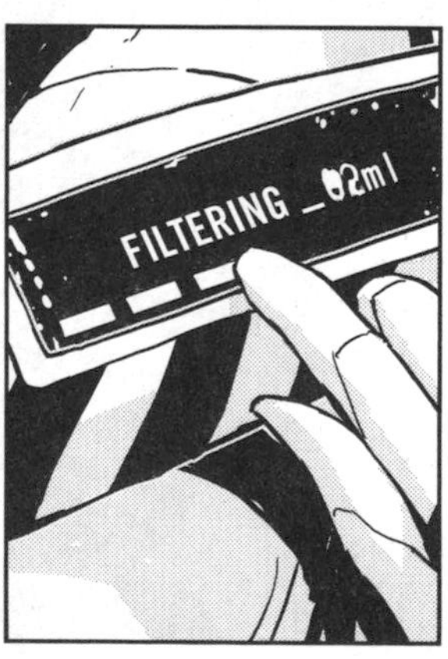
FILTERING _02ml

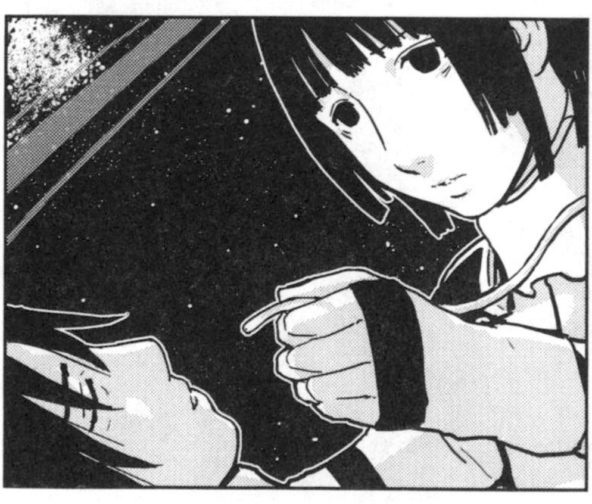

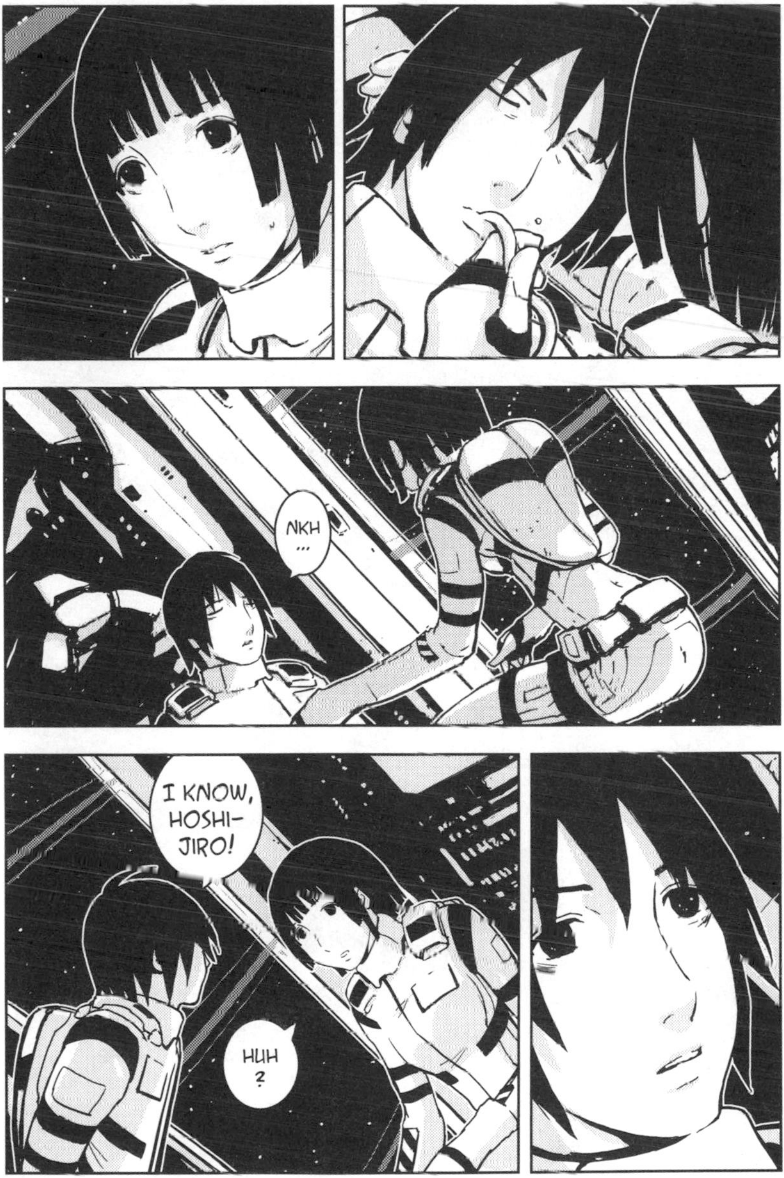
NKH ...
I KNOW, HOSHI-JIRO!
HUH?

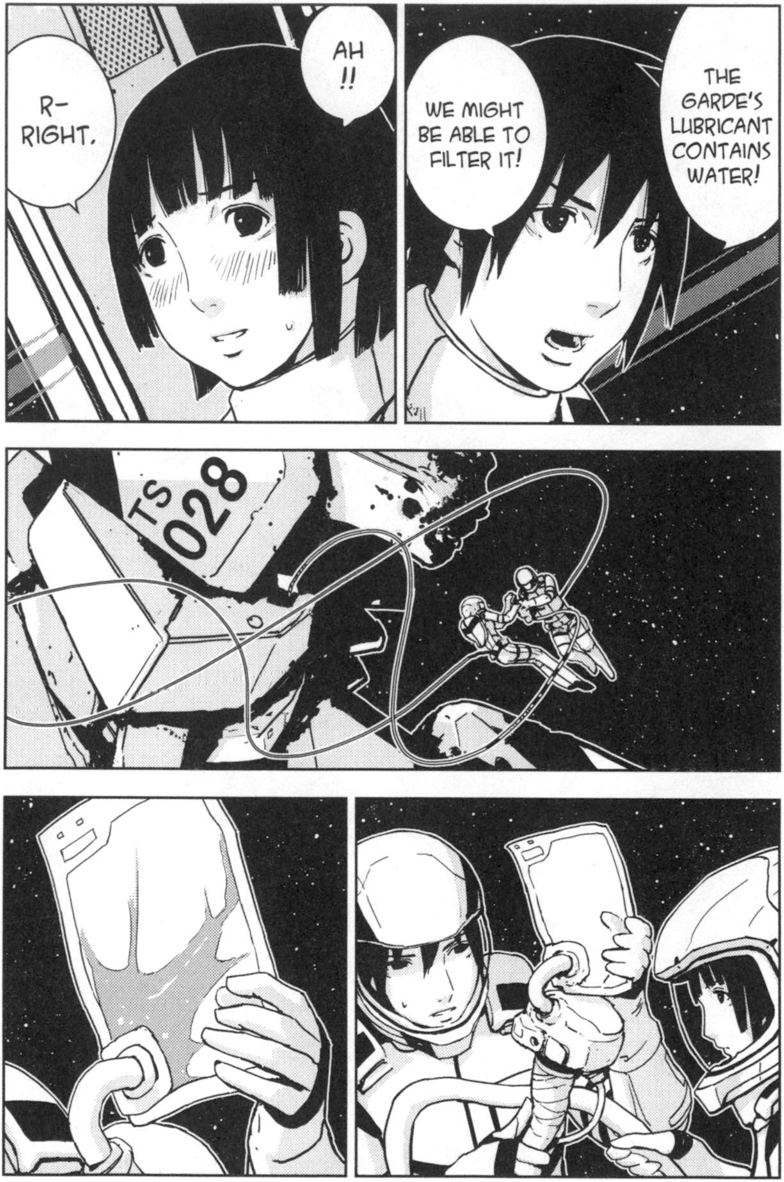
THE GARDE'S LUBRICANT CONTAINS WATER!
WE MIGHT BE ABLE TO FILTER IT!
AH !!
R-RIGHT.
TS 028

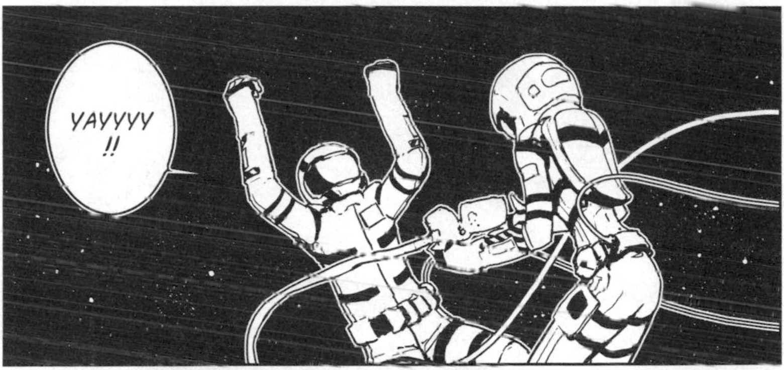
YAYYYY
!!
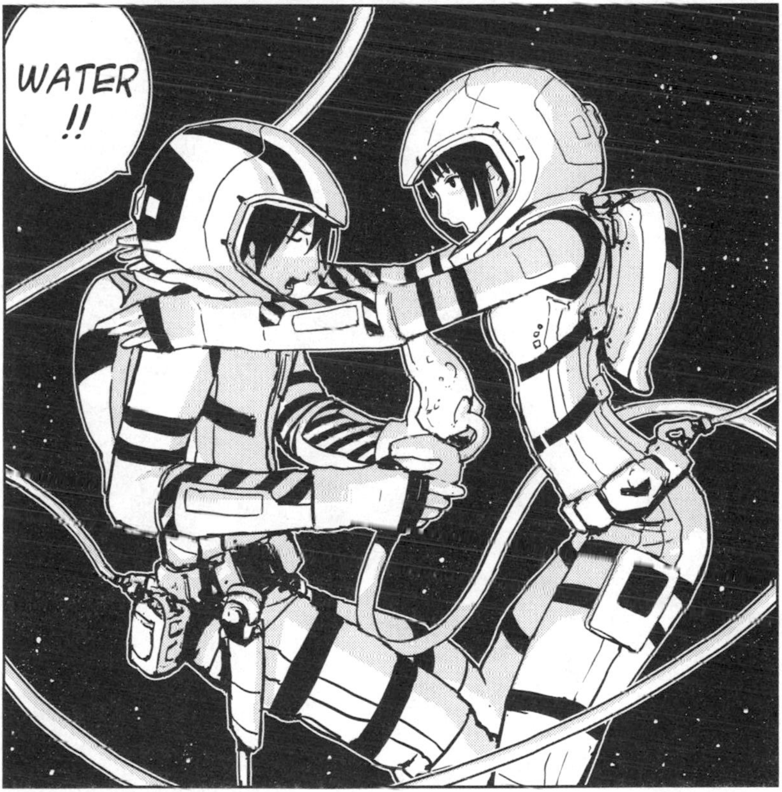
WATER
!!

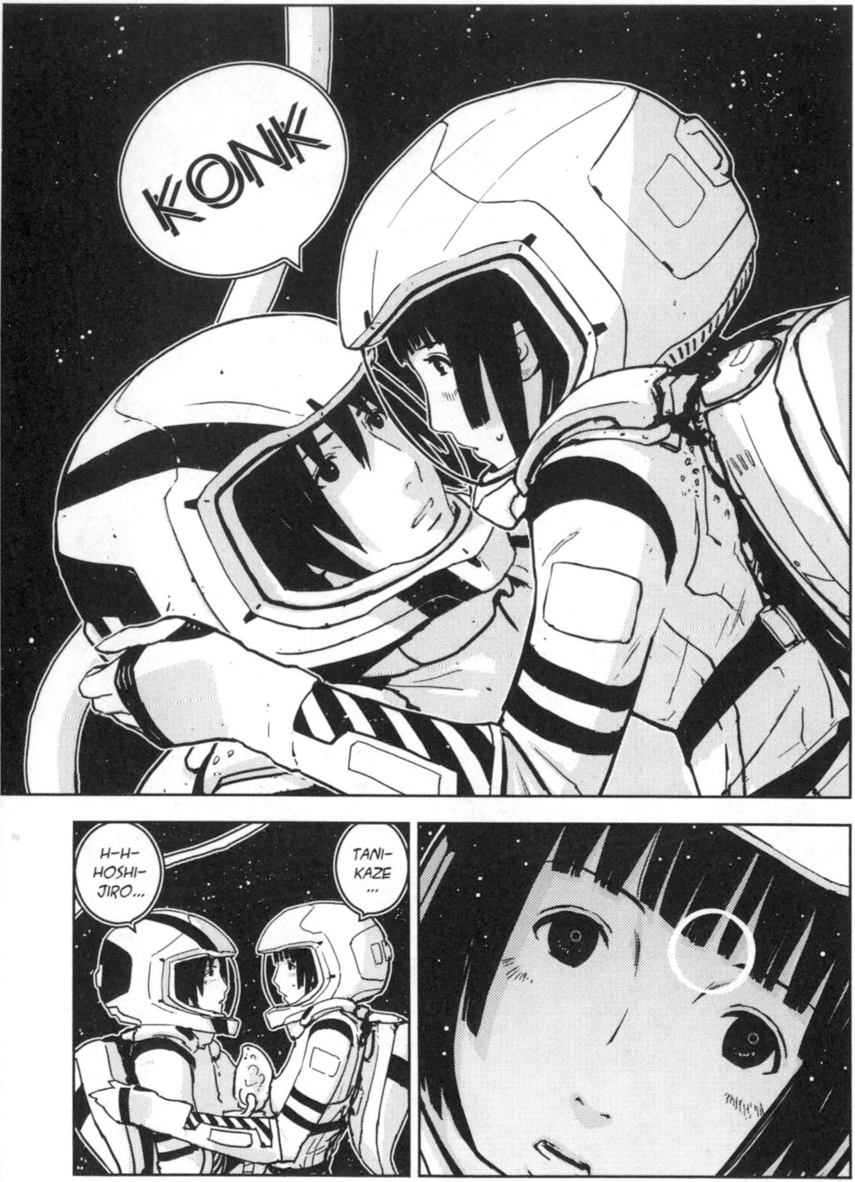
KONK
TANI-KAZE ...
H-H-HOSHI-JIRO...

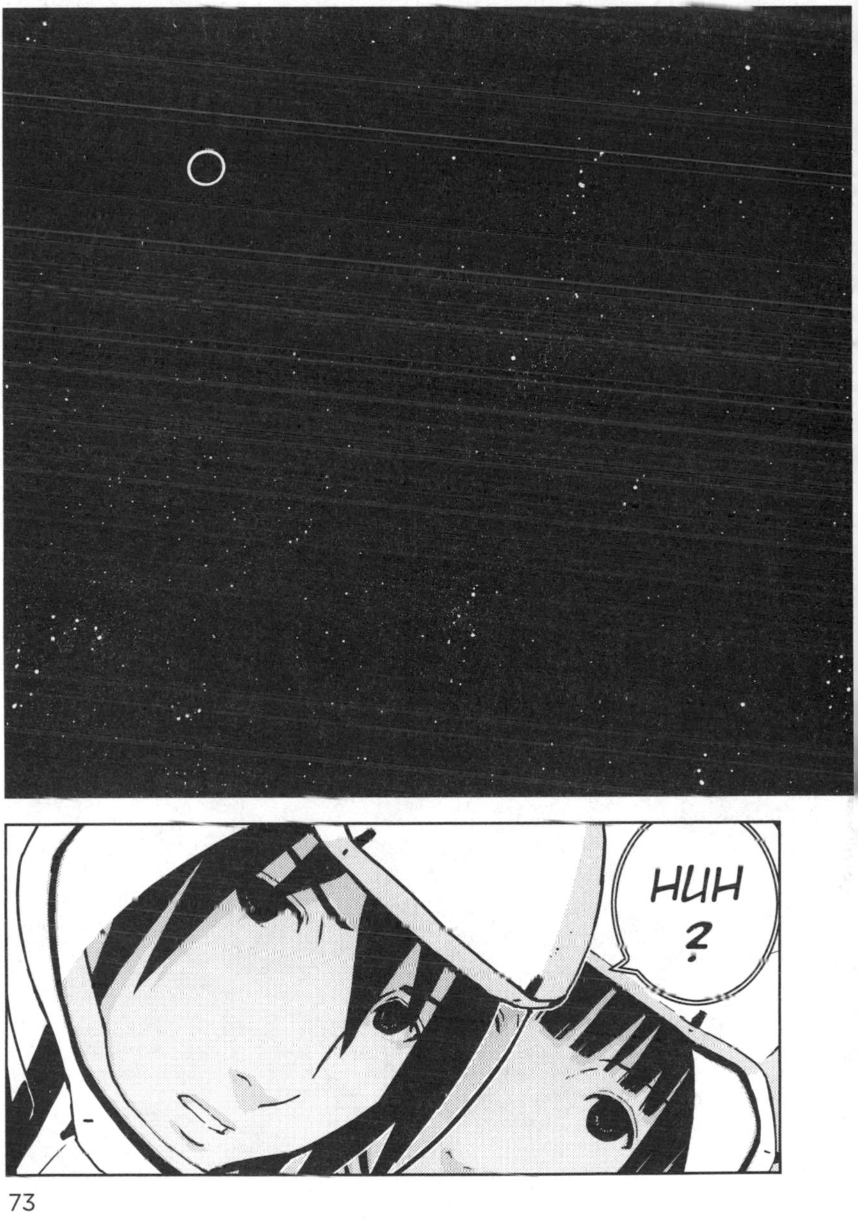
HUH?

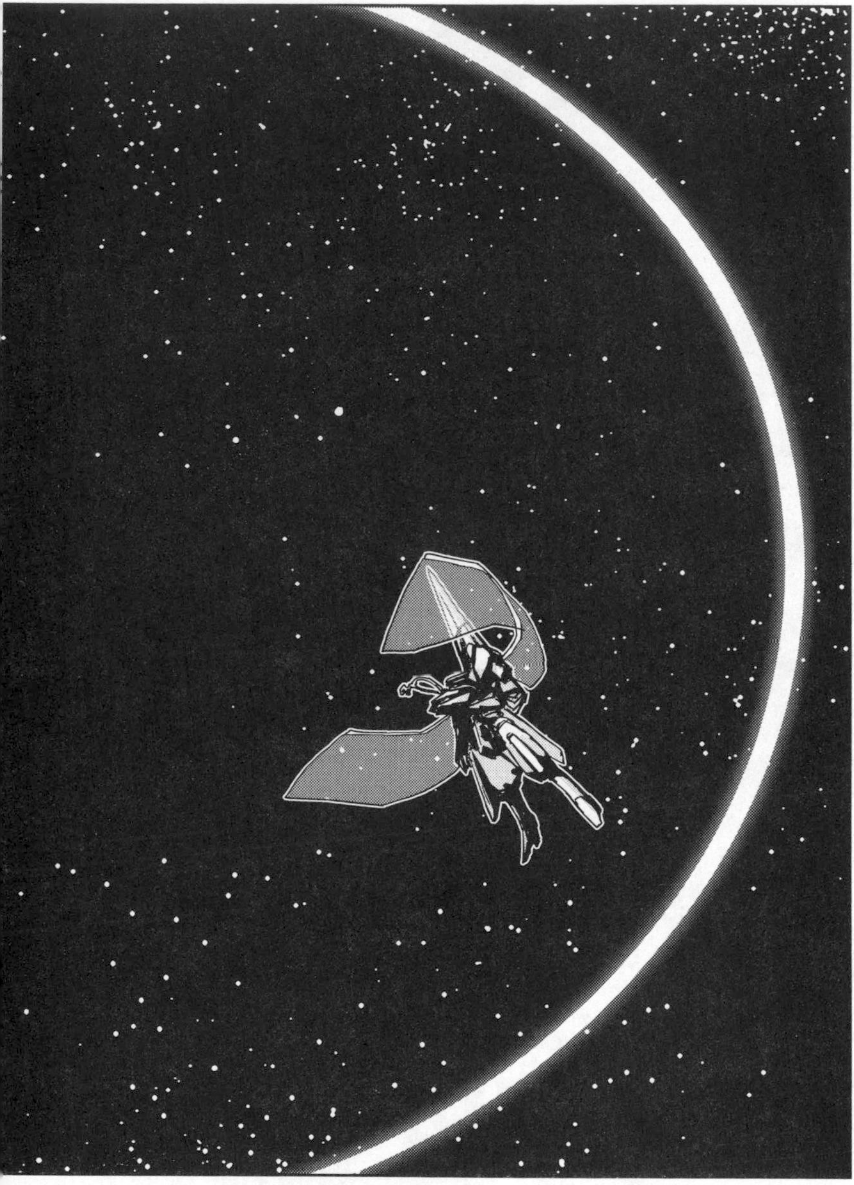

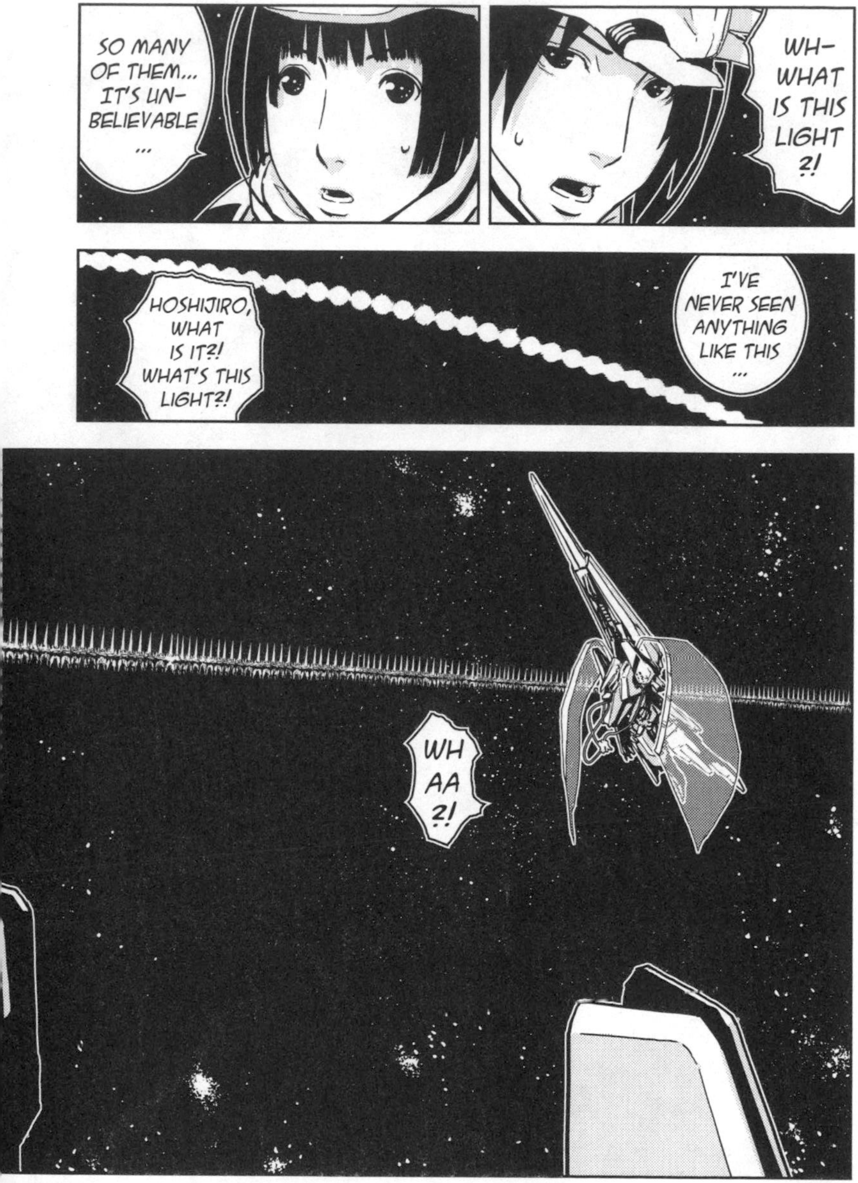
WH- WHAT IS THIS LIGHT ?!
SO MANY OF THEM... IT'S UN-BELIEVABLE ...
I'VE NEVER SEEN ANYTHING LIKE THIS ...
HOSHIJIRO, WHAT IS IT?! WHAT'S THIS LIGHT?!
WHAA ?!

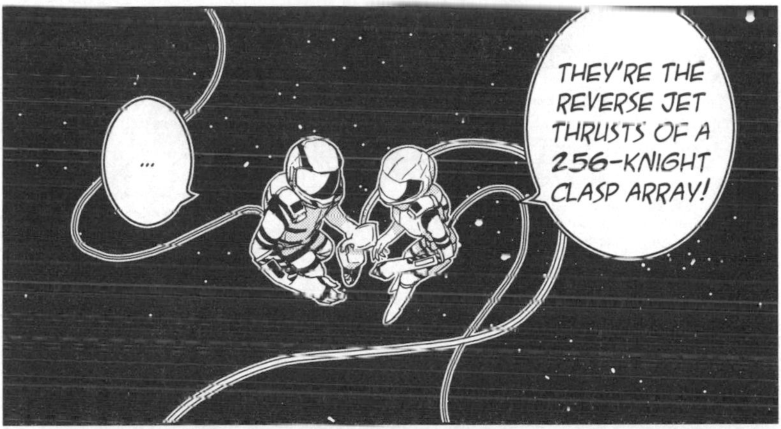
THEY'RE THE REVERSE JET THRUSTS OF A 256-KNIGHT CLASP ARRAY!
...
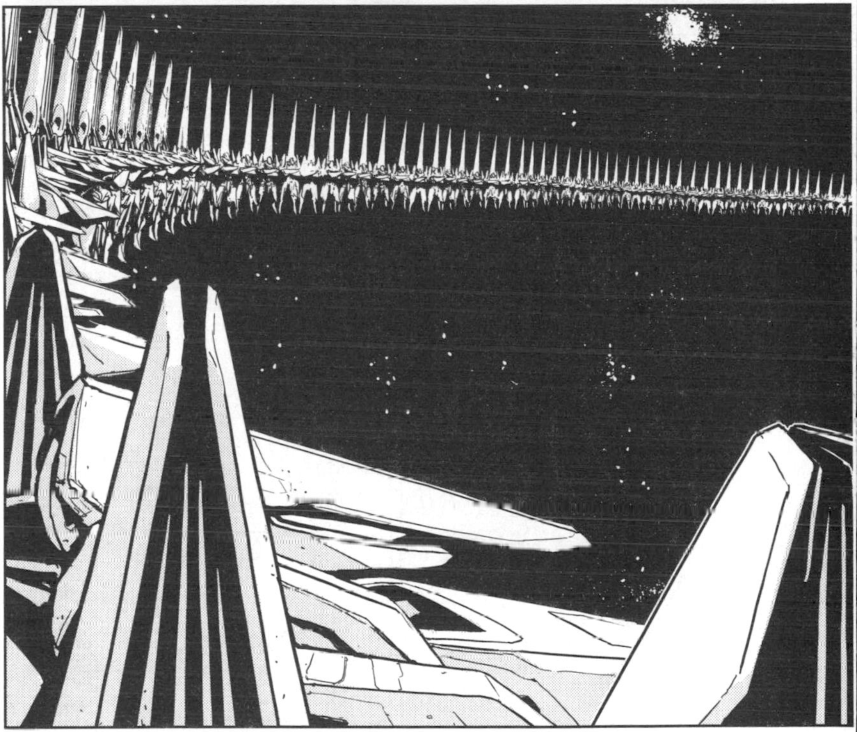

WE HAVE SECURED NAGATE TANIKAZE AND SHIZUKA HOSHIJIRO.
005
FINALLY, WE HAVE A RESPONSE!
THIS IS UNHEARD OF. THE ENTIRE GARRISON VIOLATING SHIP REGULATIONS!!
HOW DO YOU LOT PLAN TO PAY FOR THIS?
DO YOU THINK WE HAVE 257 SOLITARY CELLS?
YOU TAKE MY MEANING. OUT.
Wooh!
C-CAPTAIN...
Yahooo!
Yeah!

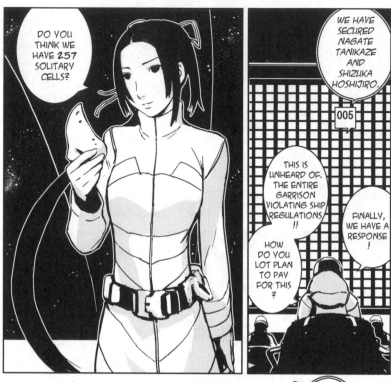

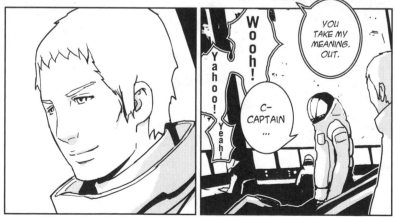

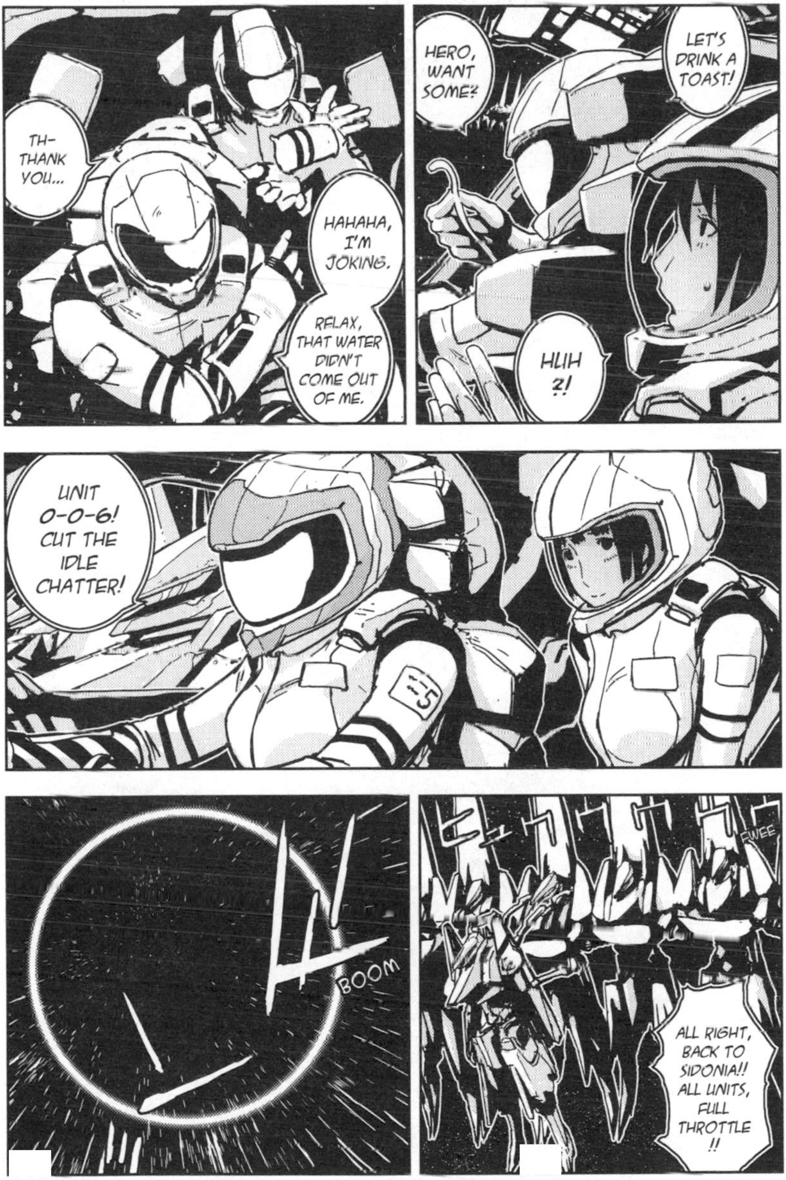
LET'S DRINK A TOAST!
HERO, WANT SOME?
HUH ?!
TH- THANK YOU...
HAHAHA, I'M JOKING.
RELAX, THAT WATER DIDN'T COME OUT OF ME.
UNIT 0-0-6! CUT THE IDLE CHATTER!
FWEE
ALL RIGHT, BACK TO SIDONIA!! ALL UNITS, FULL THROTTLE !!
BOOM

Chapter 7: END

KNIGHTS OF SIDONIA
BY TSUTOMU NIHEI

Chapter 8: Nagate Tanikaze's Return

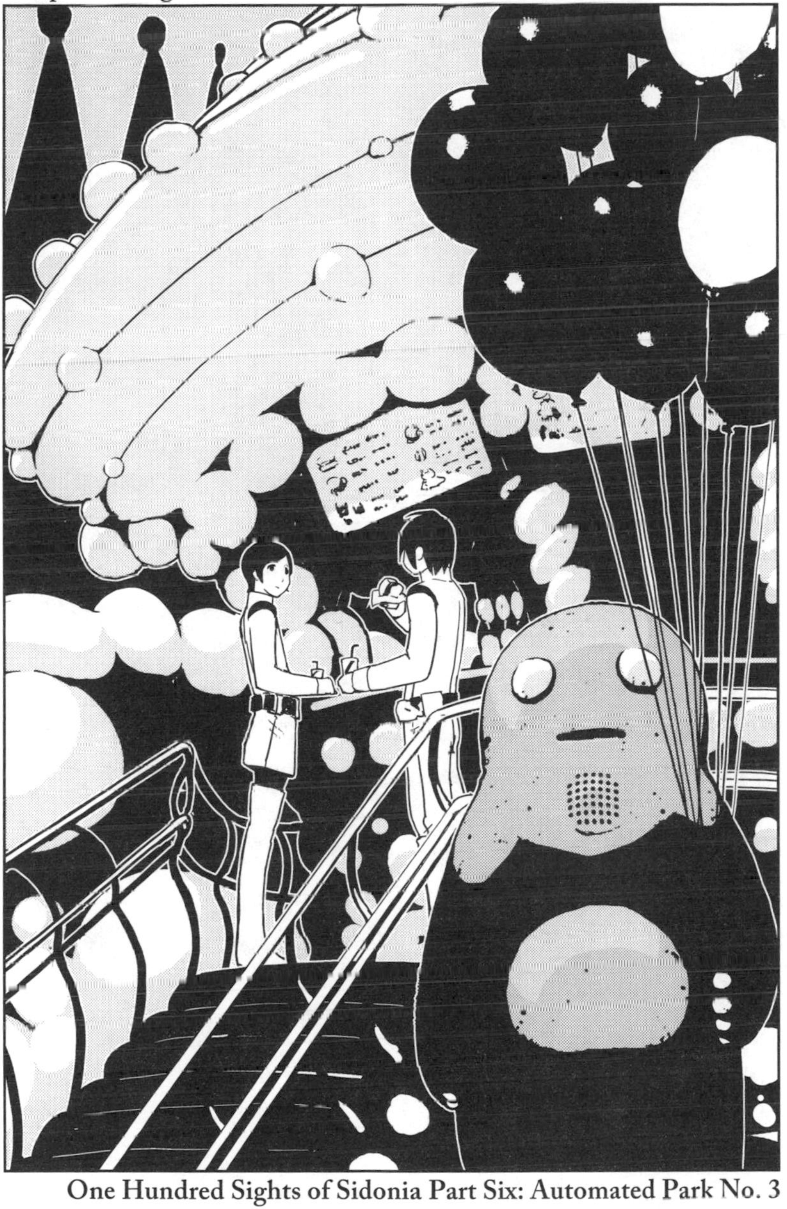

One Hundred Sights of Sidonia Part Six: Automated Park No. 3

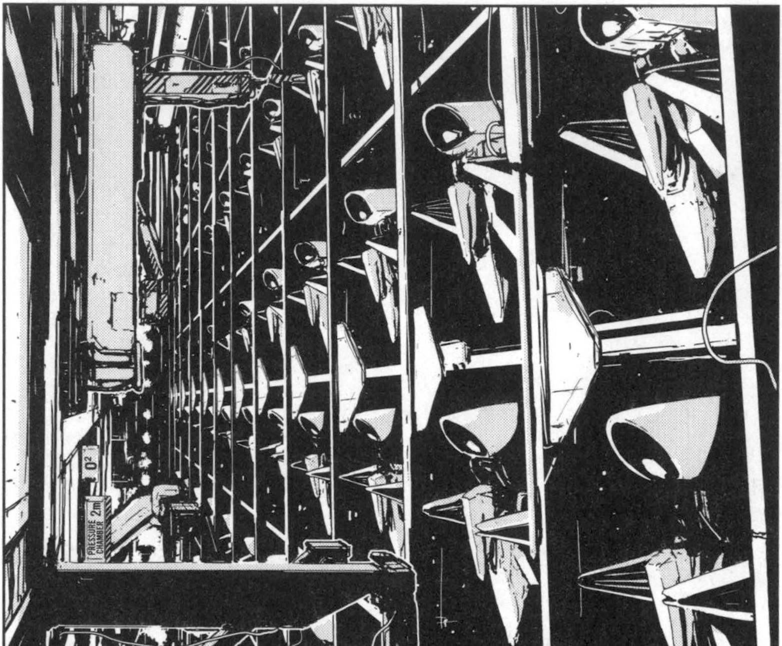
PRESSURE CHAMBER
2m
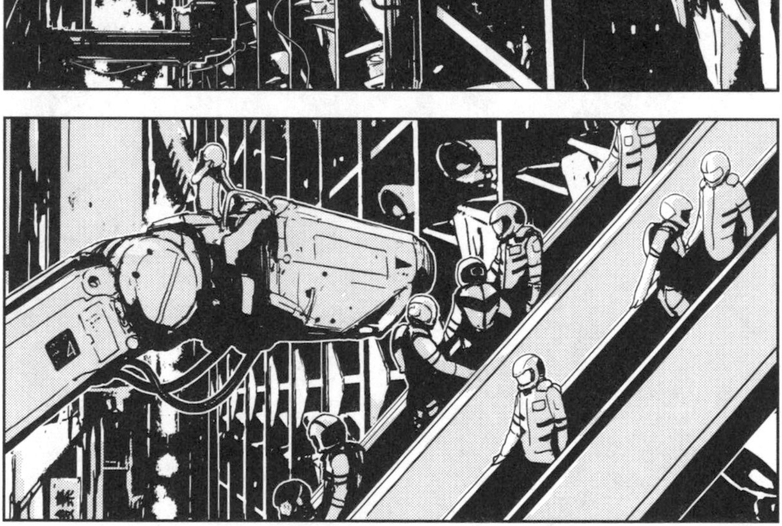

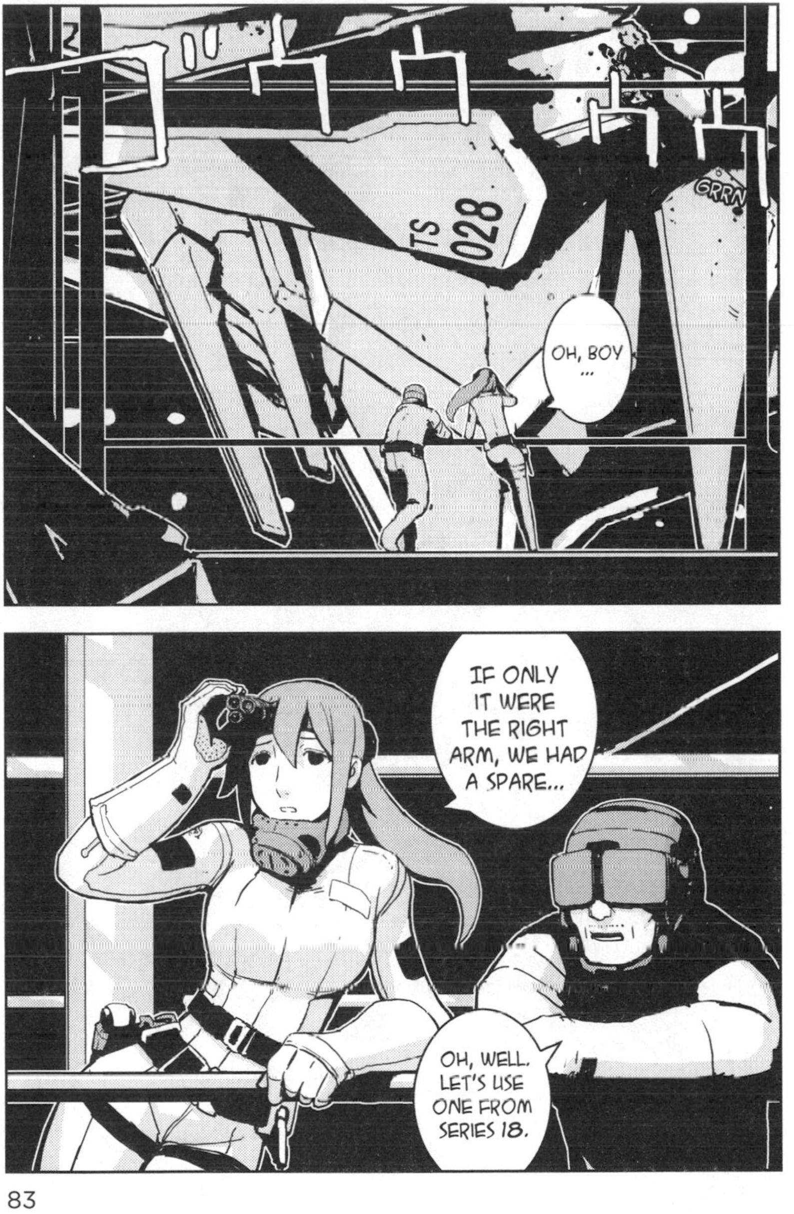
GRRN
TS 028
OH, BOY ...
IF ONLY IT WERE THE RIGHT ARM, WE HAD A SPARE...
OH, WELL. LET'S USE ONE FROM SERIES 18.

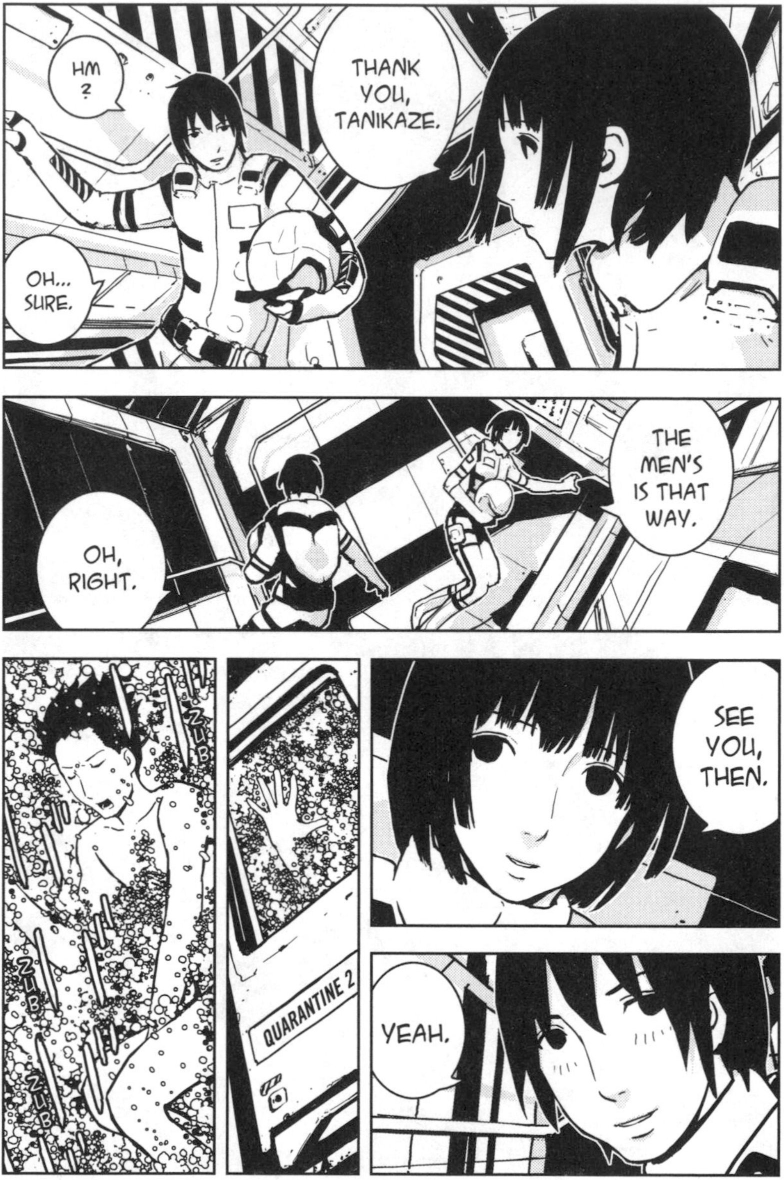
THANK YOU, TANIKAZE.
HM ?
OH... SURE.
THE MEN'S IS THAT WAY.
OH, RIGHT.
SEE YOU, THEN.
YEAH.
QUARANTINE 2
ZUB
ZUB
ZUB
ZUB

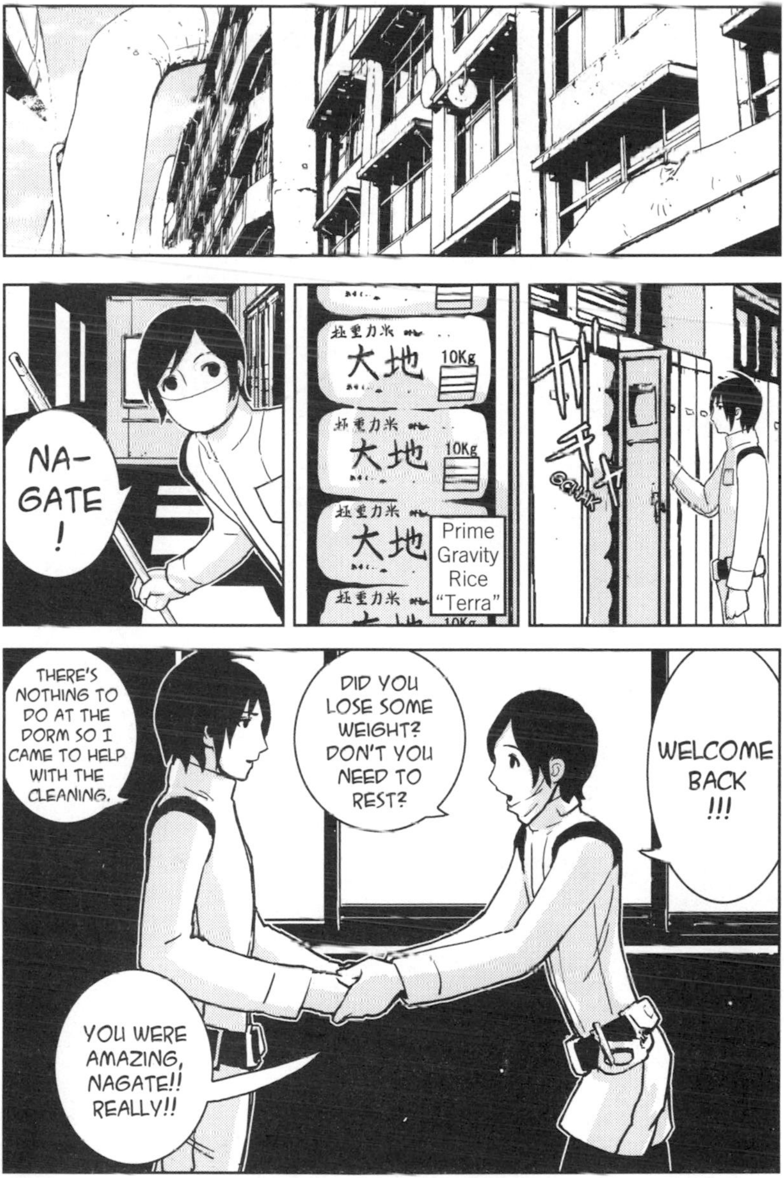

ガチャ
大地
10Kg
Prime Gravity Rice "Terra"
NA-GATE!
WELCOME BACK!!!
DID YOU LOSE SOME WEIGHT? DON'T YOU NEED TO REST?
THERE'S NOTHING TO DO AT THE DORM SO I CAME TO HELP WITH THE CLEANING.
YOU WERE AMAZING, NAGATE!! REALLY!!

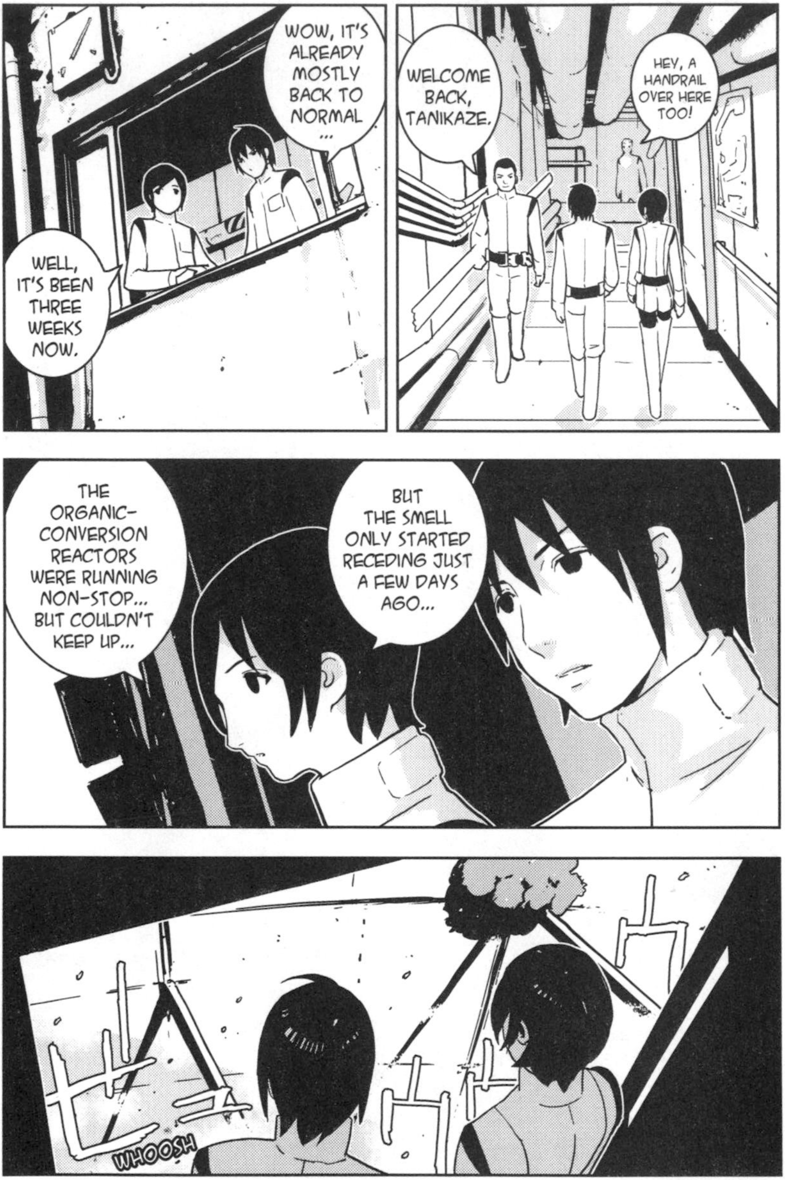
HEY, A HANDRAIL OVER HERE TOO!
WELCOME BACK, TANIKAZE.
WOW, IT'S ALREADY MOSTLY BACK TO NORMAL ...
WELL, IT'S BEEN THREE WEEKS NOW.
BUT THE SMELL ONLY STARTED RECEDING JUST A FEW DAYS AGO...
THE ORGANIC-CONVERSION REACTORS WERE RUNNING NON-STOP... BUT COULDN'T KEEP UP...
WHOOSH

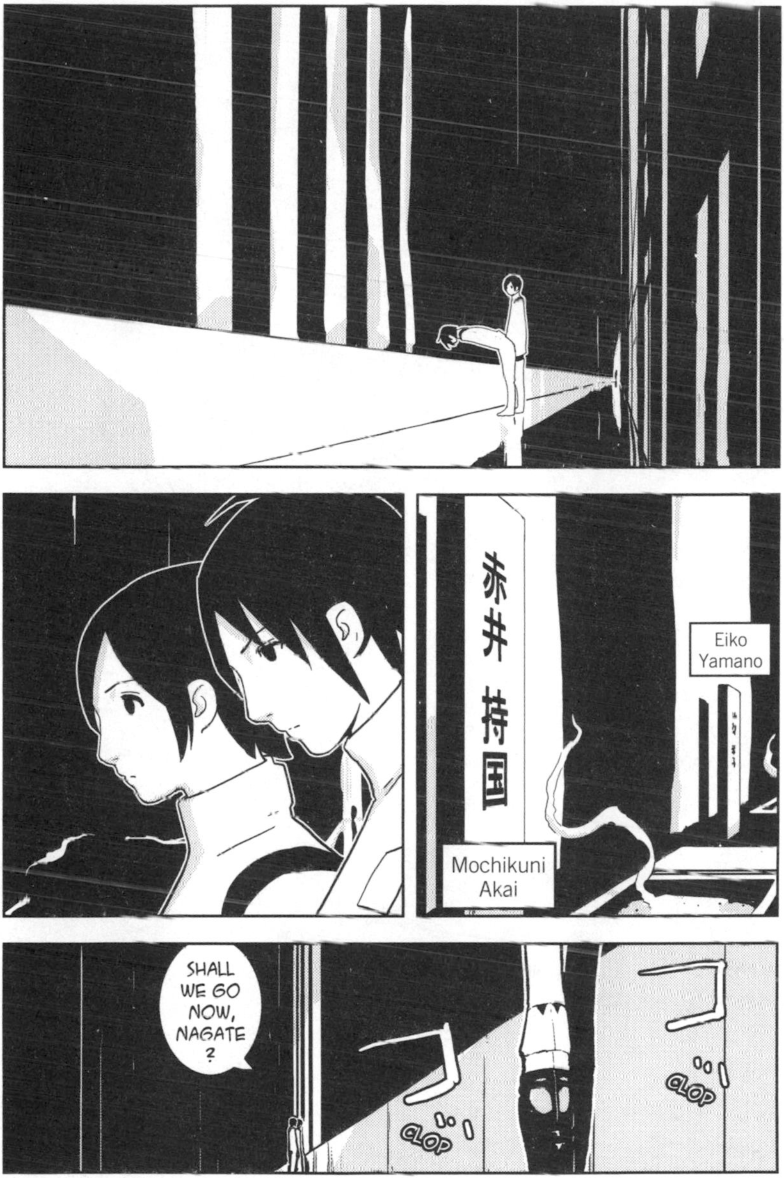
Eiko Yamano
赤井持国
Mochikuni Akai
SHALL WE GO NOW, NAGATE?
CLOP
CLOP

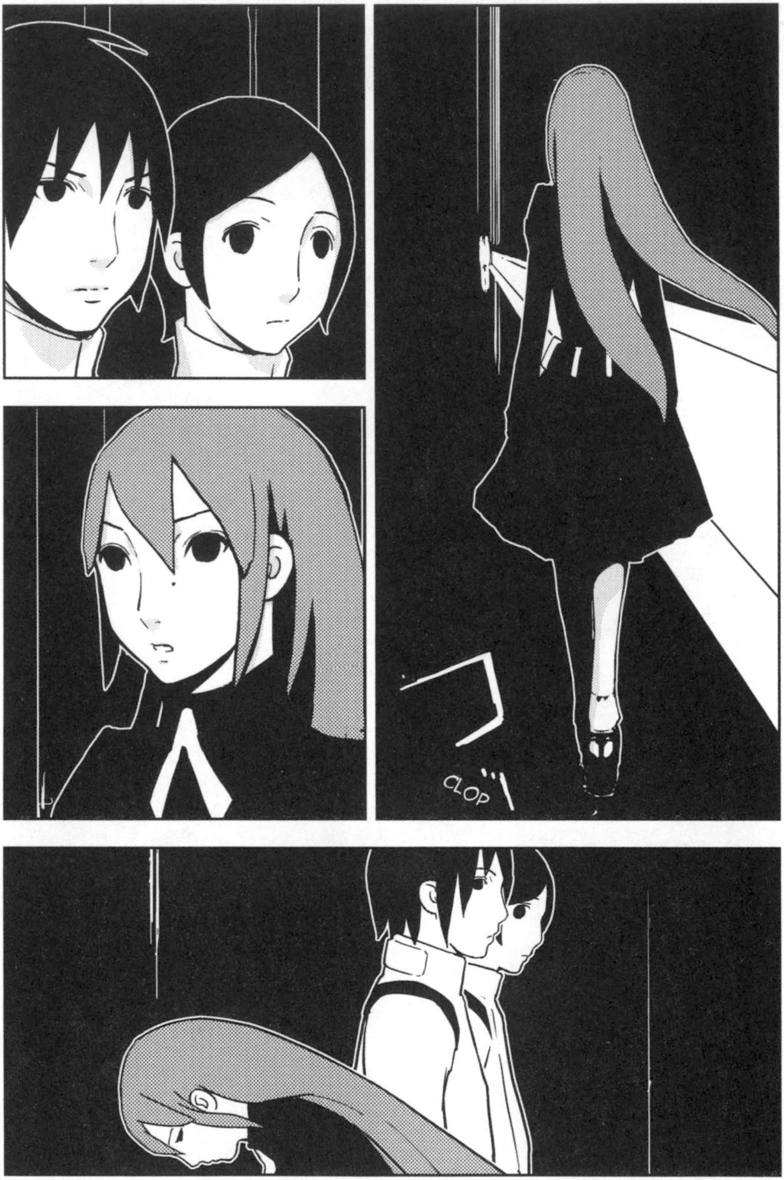
CLOP

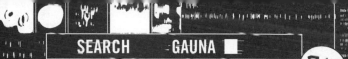

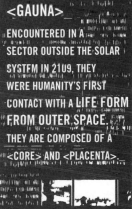

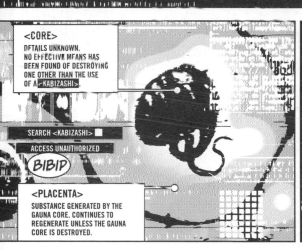

SEARCH GAUNA
BIP
<GAUNA>
ENCOUNTERED IN A SECTOR OUTSIDE THE SOLAR SYSTEM IN 2109, THEY WERE HUMANITY'S FIRST CONTACT WITH A LIFE FORM FROM OUTER SPACE. THEY ARE COMPOSED OF A <CORE> AND <PLACENTA>.
<CORE>
DETAILS UNKNOWN. NO EFFECTIVE MEANS HAS BEEN FOUND OF DESTROYING ONE OTHER THAN THE USE OF A <KABIZASHI>.
SEARCH <KABIZASHI>
ACCESS UNAUTHORIZED
BIBIP
<PLACENTA>
SUBSTANCE GENERATED BY THE GAUNA CORE. CONTINUES TO REGENERATE UNLESS THE GAUNA CORE IS DESTROYED.
<CLUSTER SHIP>
AN AGGREGATE BODY OF NUMEROUS GAUNAS. THE FIRST <CLUSTER SHIP> ENCOUNTERED BY HUMANITY HAD A DIAMETER OF APPROX. 5000 KILOMETERS.
CAME KNOWN THAT IT HAS A POMEGRANATE-LIKE INTERNAL ORGANIZATION. THE ESTIMATED NUMBER OF GAUNAS CONTAINED: SEVERAL TENS OF BILLIONS.
2371
MORE THAN TWO HUNDRED YEARS AFTER THE FIRST ENCOUNTER, AND WITH ABSOLUTELY NO RESPONSE TO HUMANITY'S ATTEMPTED COMMUNICATIONS, THE <CLUSTER SHIP> SUDDENLY APPROACHED THE SOLAR SYSTEM.
FORTY-SIX <GAUNAS> WERE DROPPED ONTO EARTH.

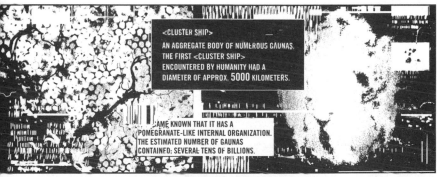

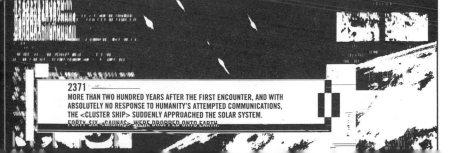

THE FIRST GAUNA DROPPED ON EARTH

KSHKK

A GAUNA PREYING UPON MARINE LIFE

KSHKK

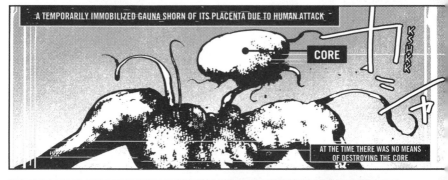

A TEMPORARILY IMMOBILIZED GAUNA SHORN OF ITS PLACENTA DUE TO HUMAN ATTACK

CORE

AT THE TIME THERE WAS NO MEANS
OF DESTROYING THE CORE

KSHKK

THE MOMENT OF
EARTH'S BISECTION

KSHKK

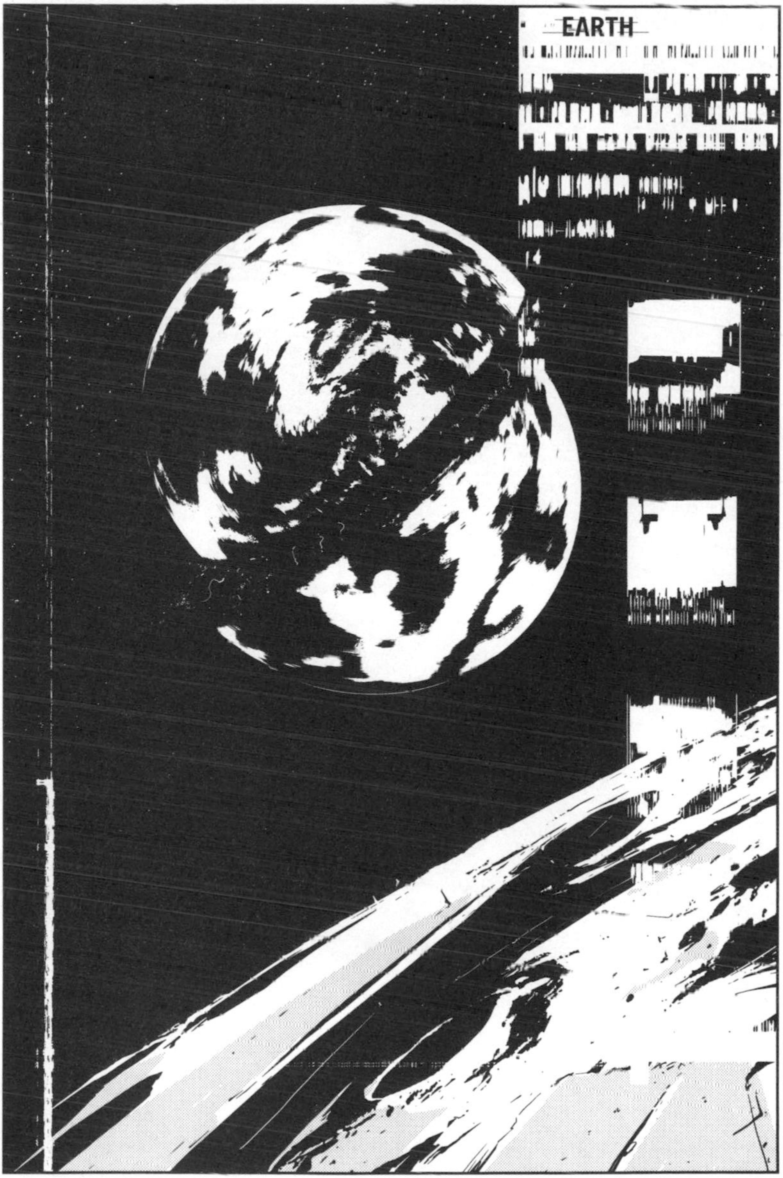
EARTH

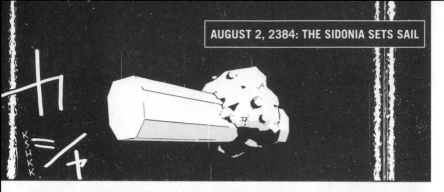

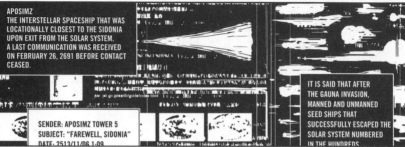

AUGUST 2, 2384: THE SIDONIA SETS SAIL
KSHKK
IT IS SAID THAT AFTER THE GAUNA INVASION, MANNED AND UNMANNED SEED SHIPS THAT SUCCESSFULLY ESCAPED THE SOLAR SYSTEM NUMBERED IN THE HUNDREDS.
APOSIMZ
THE INTERSTELLAR SPACESHIP THAT WAS LOCATIONALLY CLOSEST TO THE SIDONIA UPON EXIT FROM THE SOLAR SYSTEM. A LAST COMMUNICATION WAS RECEIVED ON FEBRUARY 26, 2691 BEFORE CONTACT CEASED.
SENDER: APOSIMZ TOWER 5
SUBJECT: "FAREWELL, SIDONIA"
DATE: 2513/11/06 1:09
AT PRESENT, NOVEMBER 25, 1009 SIDONIA DEPARTURE ERA (COMMON ERA 3394), THERE IS NO COMMUNICATION WITH OTHER SHIPS.
GROWL LLLLL

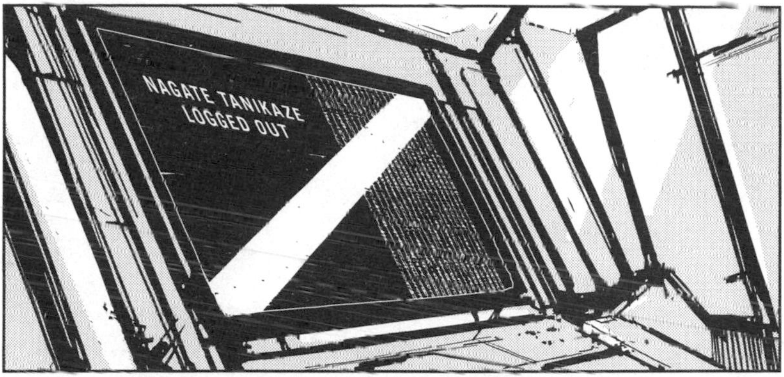
NAGATE TANIKAZE
LOGGED OUT

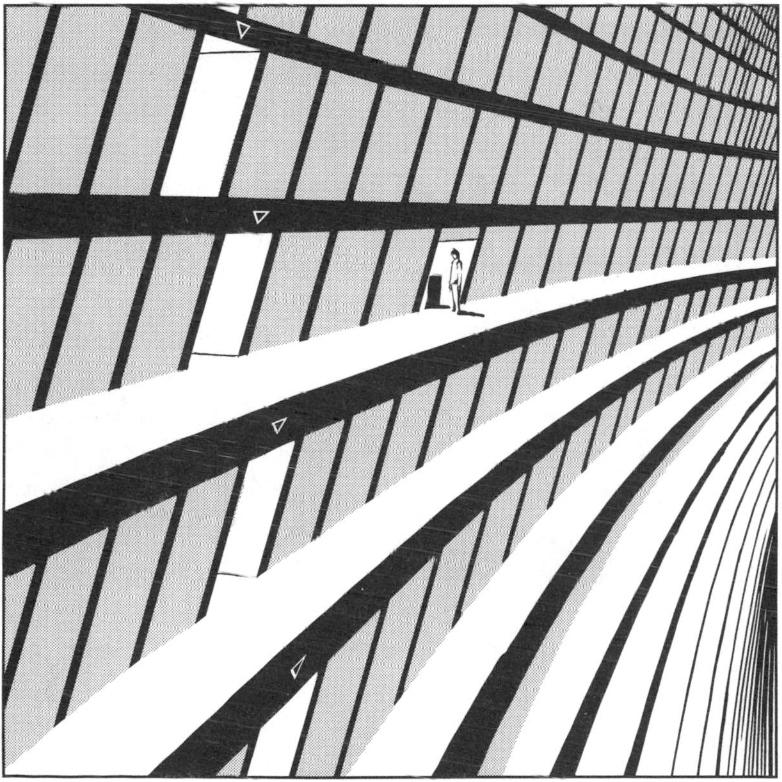

NORIO KUNATO. SHIZUKA HOSHIJIRO. EN HONOKA. NAGATE TANIKAZE.

THESE FOUR ARE HEREBY APPOINTED AS GARDE PILOTS.

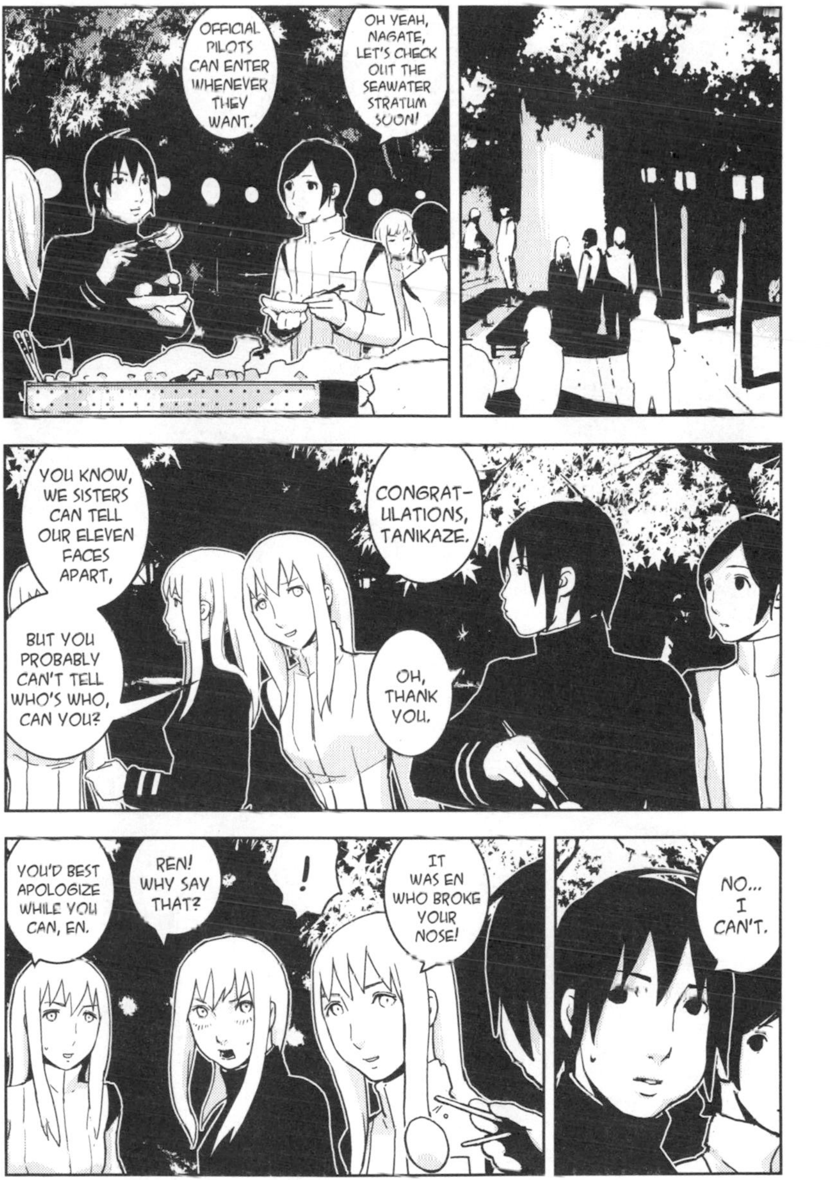
OH YEAH, NAGATE, LET'S CHECK OUT THE SEAWATER STRATUM SOON!
OFFICIAL PILOTS CAN ENTER WHENEVER THEY WANT.
CONGRAT-ULATIONS, TANIKAZE.
OH, THANK YOU.
YOU KNOW, WE SISTERS CAN TELL OUR ELEVEN FACES APART,
BUT YOU PROBABLY CAN'T TELL WHO'S WHO, CAN YOU?
NO... I CAN'T.
IT WAS EN WHO BROKE YOUR NOSE!
!
REN! WHY SAY THAT?
YOU'D BEST APOLOGIZE WHILE YOU CAN, EN.

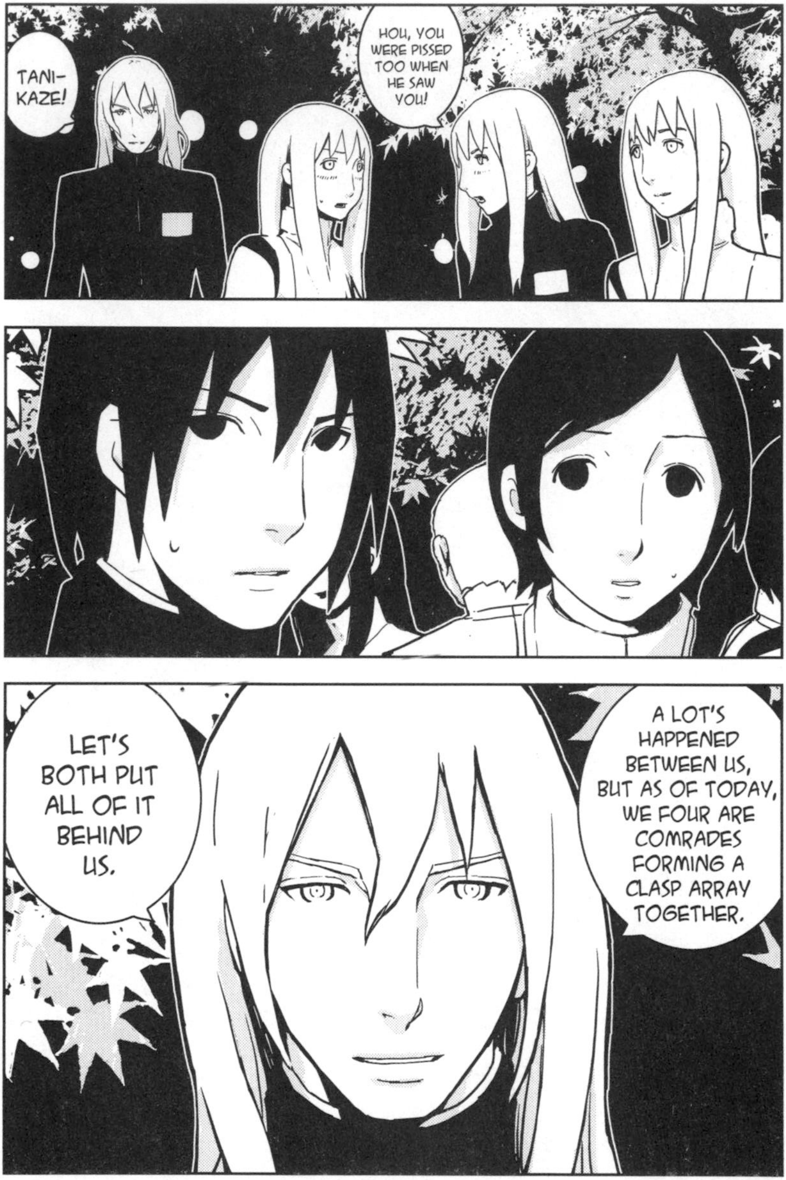
HOU, YOU WERE PISSED TOO WHEN HE SAW YOU!
TANI-KAZE!
A LOT'S HAPPENED BETWEEN US, BUT AS OF TODAY, WE FOUR ARE COMRADES FORMING A CLASP ARRAY TOGETHER.
LET'S BOTH PUT ALL OF IT BEHIND US.

HOSHI-JIRO!

HOW ABOUT A CLASP RIGHT NOW, AS A QUARTET?

YEAH, OF COURSE.

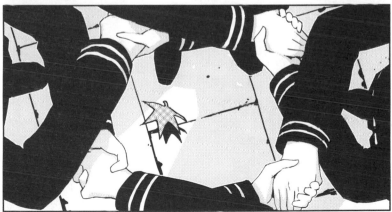

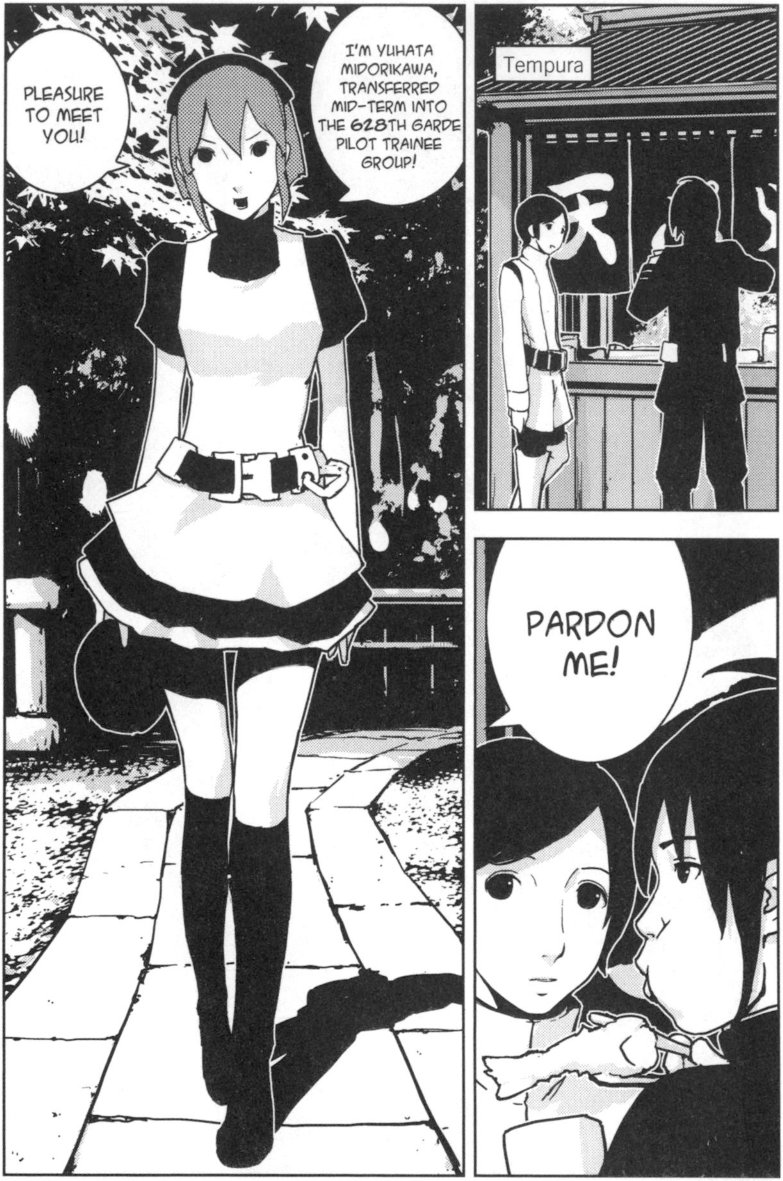
Tempura
PARDON ME!
I'M YUHATA MIDORIKAWA, TRANSFERRED MID-TERM INTO THE 628TH GARDE PILOT TRAINEE GROUP!
PLEASURE TO MEET YOU!

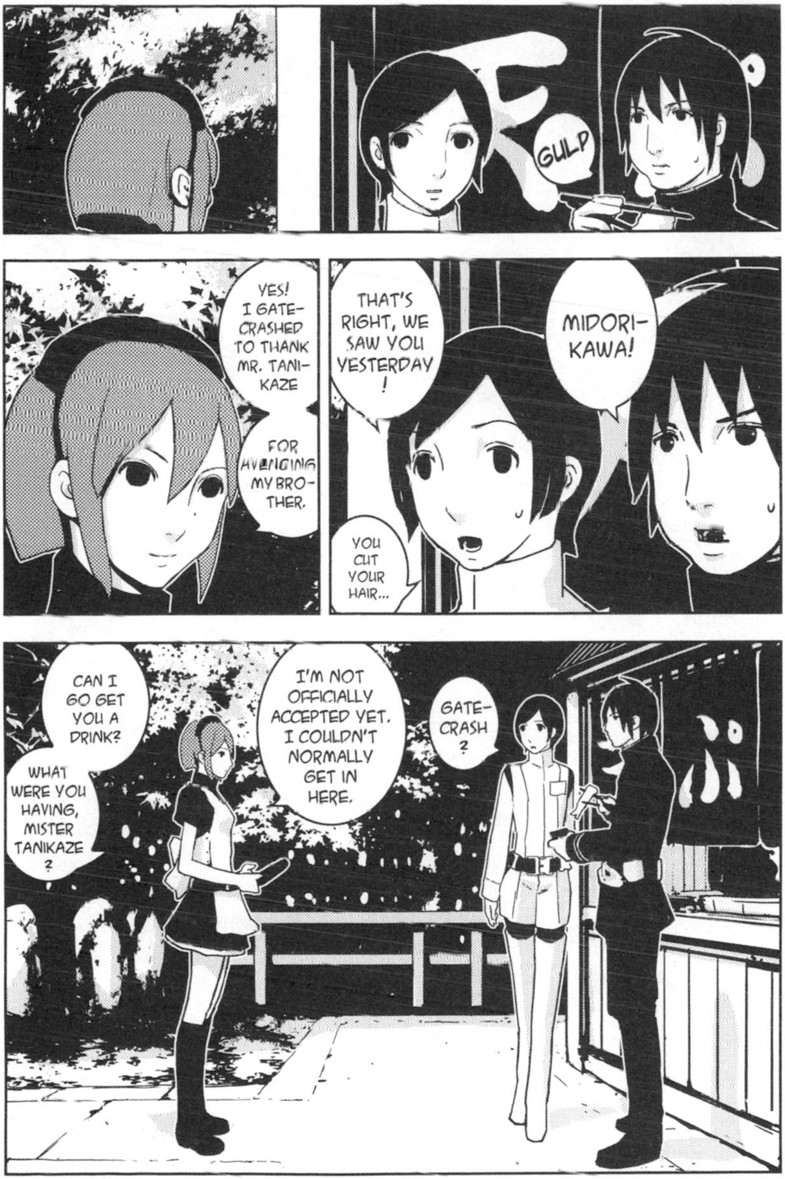
GULP
MIDORI-KAWA!
THAT'S RIGHT, WE SAW YOU YESTERDAY!
YOU CUT YOUR HAIR...
YES! I GATE-CRASHED TO THANK MR. TANI-KAZE
FOR AVENGING MY BRO-THER.
GATE-CRASH?
I'M NOT OFFICIALLY ACCEPTED YET. I COULDN'T NORMALLY GET IN HERE.
CAN I GO GET YOU A DRINK?
WHAT WERE YOU HAVING, MISTER TANIKAZE?

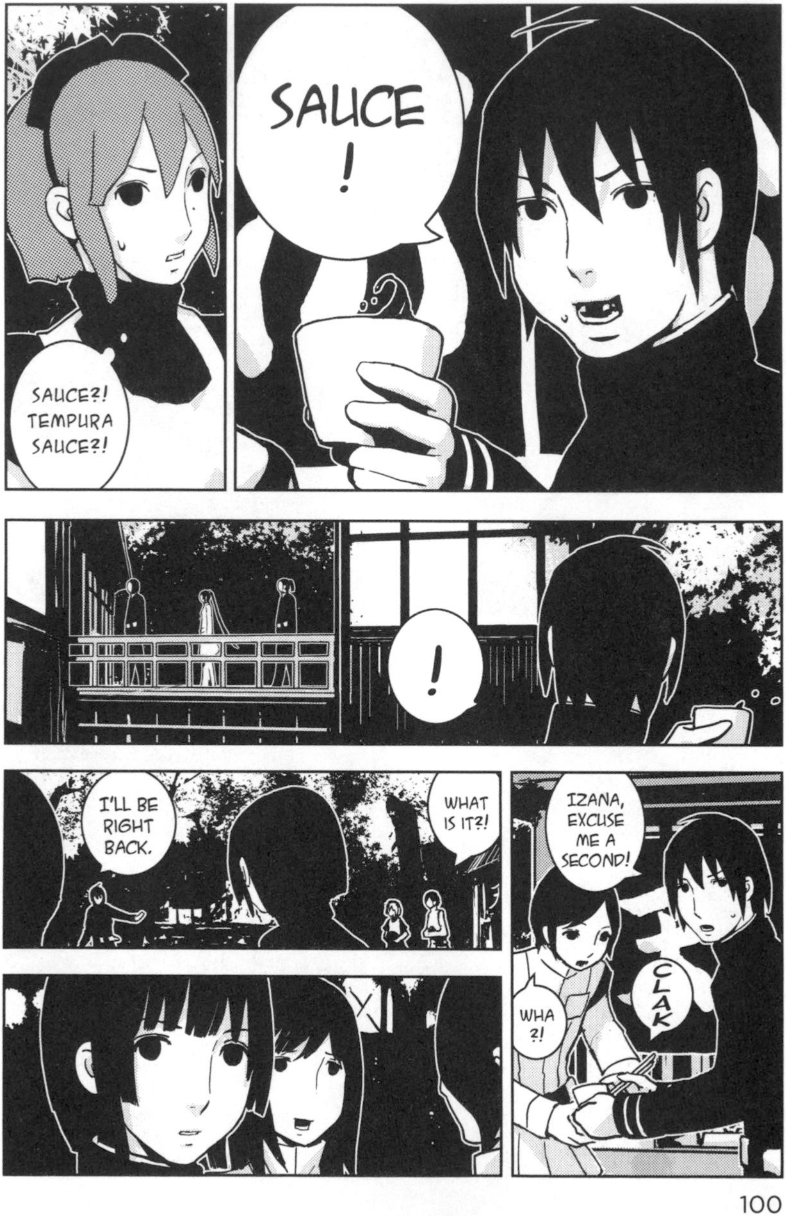
SAUCE !
SAUCE?! TEMPURA SAUCE?!
!
IZANA, EXCUSE ME A SECOND!
CLAK
WHA?!
WHAT IS IT?!
I'LL BE RIGHT BACK.

WHO ARE YOU.

...IT'S FINE, LET HIM THROUGH.

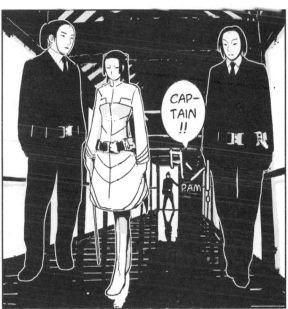

CAPTAIN!!

P.AM

!

GO ON.

GET IN.

GWRR

GRKN

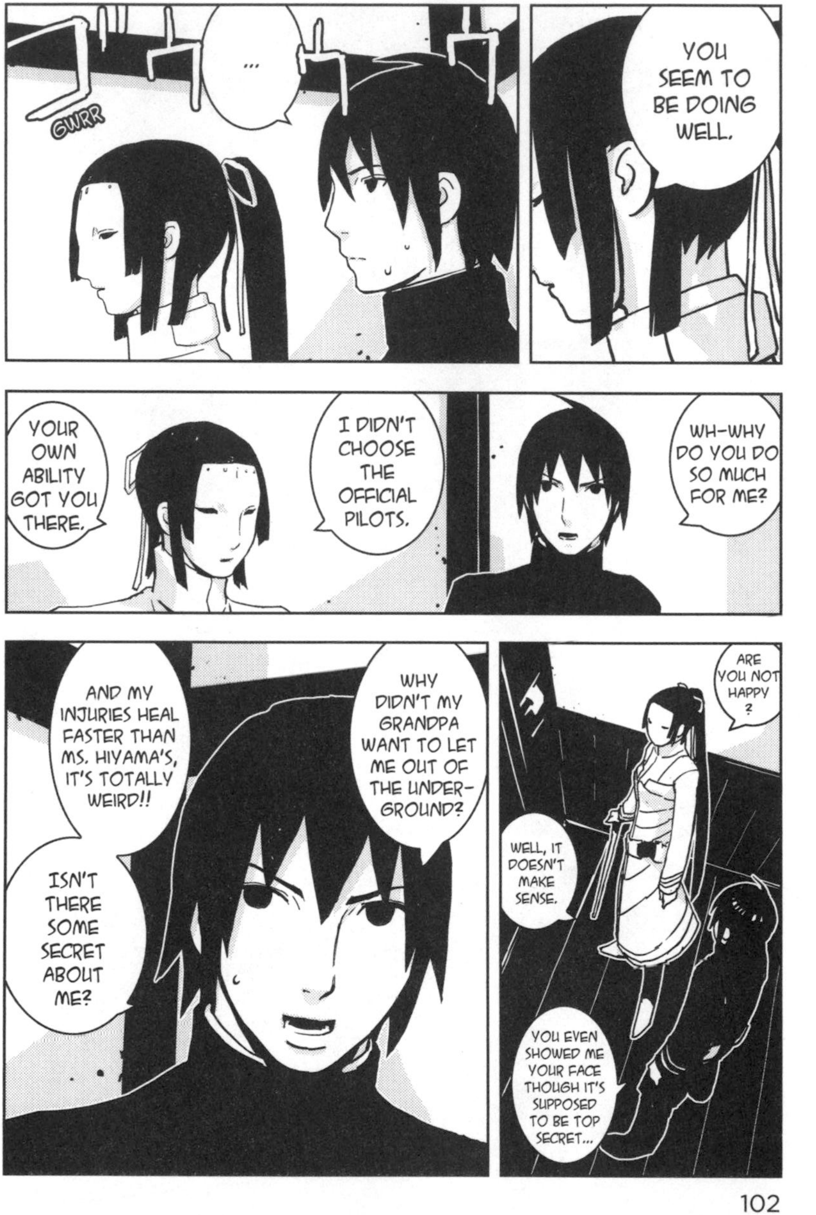
YOU SEEM TO BE DOING WELL.
...
GWRR
WH-WHY DO YOU DO SO MUCH FOR ME?
I DIDN'T CHOOSE THE OFFICIAL PILOTS.
YOUR OWN ABILITY GOT YOU THERE.
ARE YOU NOT HAPPY?
WELL, IT DOESN'T MAKE SENSE.
YOU EVEN SHOWED ME YOUR FACE THOUGH IT'S SUPPOSED TO BE TOP SECRET...
WHY DIDN'T MY GRANDPA WANT TO LET ME OUT OF THE UNDER-GROUND?
AND MY INJURIES HEAL FASTER THAN MS. HIYAMA'S, IT'S TOTALLY WEIRD!!
ISN'T THERE SOME SECRET ABOUT ME?

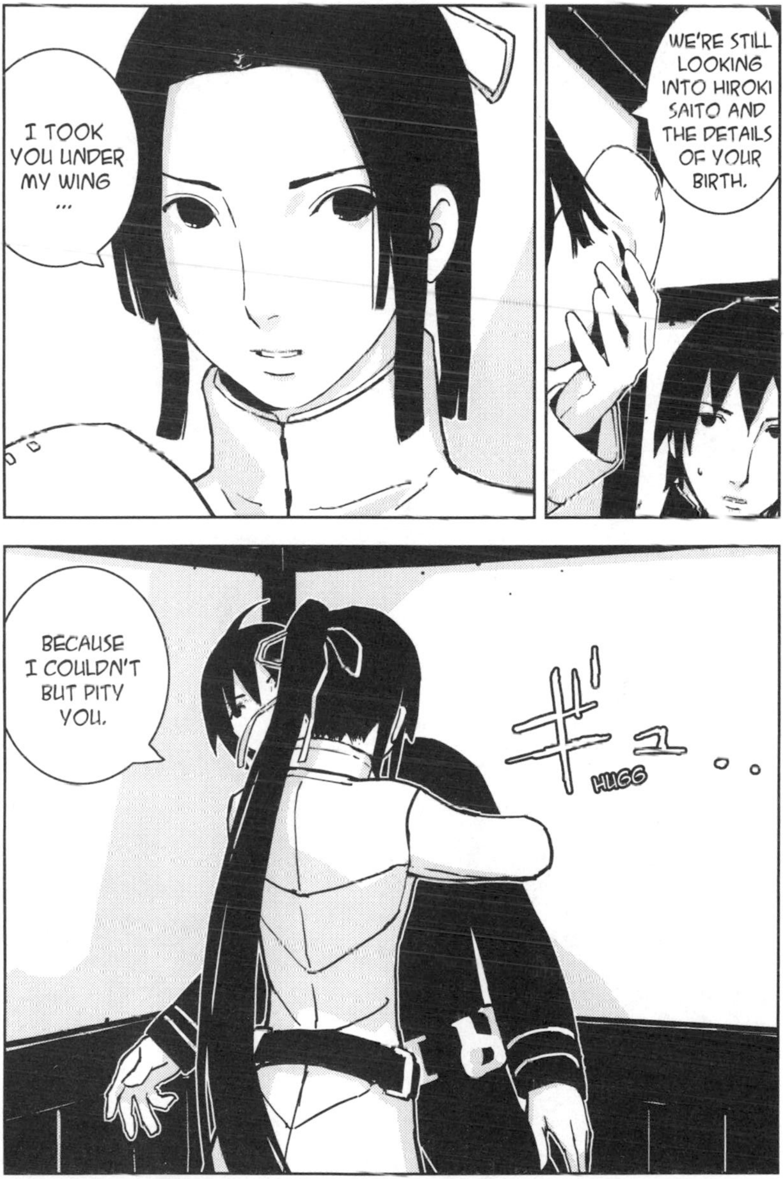
WE'RE STILL LOOKING INTO HIROKI SAITO AND THE DETAILS OF YOUR BIRTH.
I TOOK YOU UNDER MY WING ...
BECAUSE I COULDN'T BUT PITY YOU.
ギュ。。
HUGG

EVERYTHING ALL RIGHT? WERE YOU CHASING AFTER THE CAPTAIN?

HOSHI-JIRO.

!

OH... GOOD.

I JUST WANTED TO PAY MY REGARDS ...

UH, YEAH.

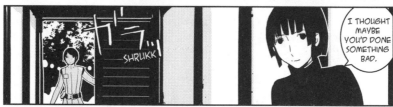

SHRUKK

I THOUGHT MAYBE YOU'D DONE SOMETHING BAD.

NAGATE, LET'S GO RIDE THAT UNDERWATER FLOATING TANK.

WHAT? NOW?!

OH...

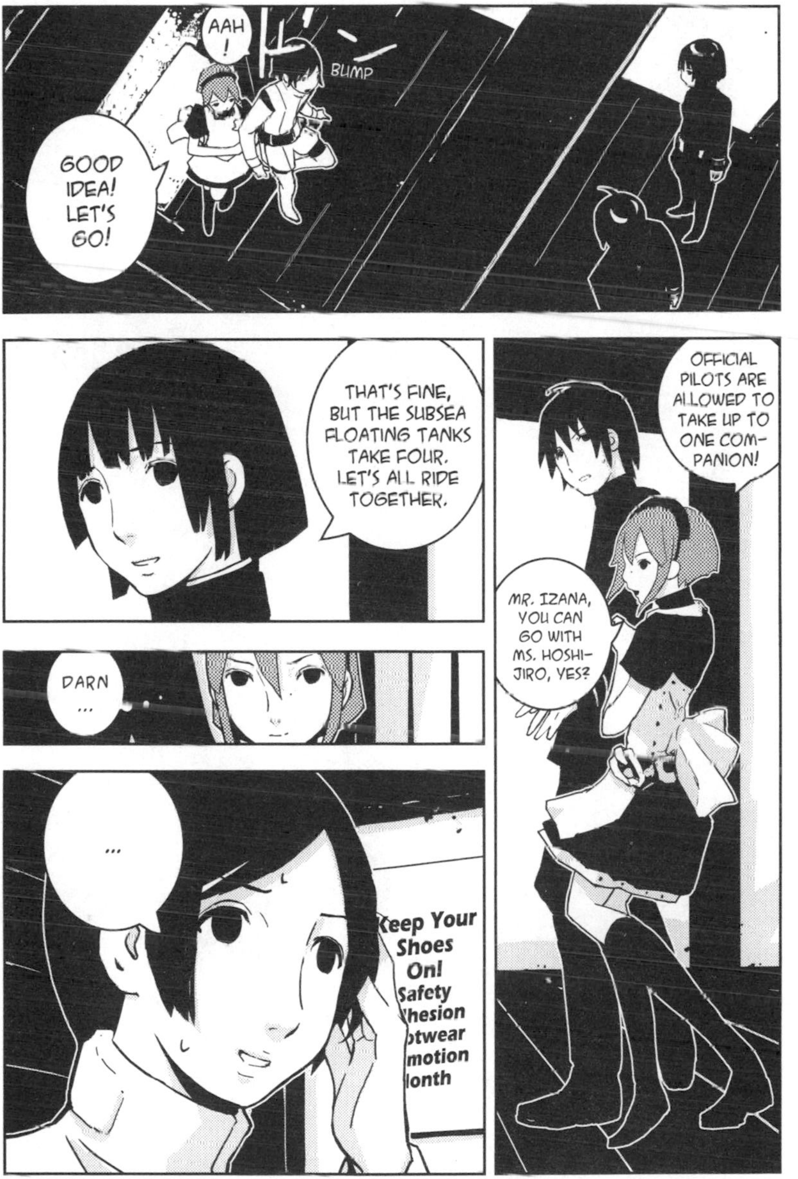
AAH !
BUMP
GOOD IDEA! LET'S GO!
OFFICIAL PILOTS ARE ALLOWED TO TAKE UP TO ONE COM-PANION!
MR. IZANA, YOU CAN GO WITH MS. HOSHI-JIRO, YES?
THAT'S FINE, BUT THE SUBSEA FLOATING TANKS TAKE FOUR. LET'S ALL RIDE TOGETHER.
DARN ...
...
Keep Your Shoes On!
Safety

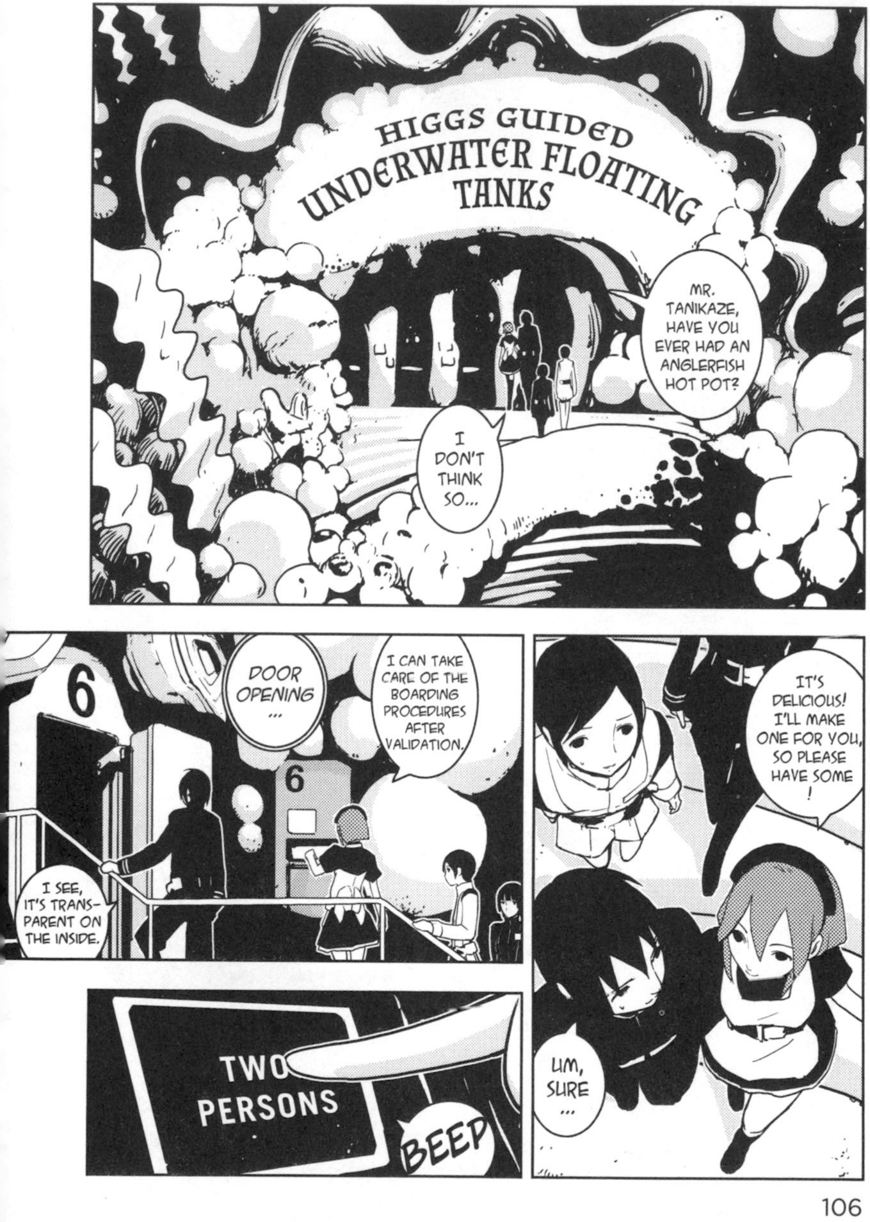
HIGGS GUIDED UNDERWATER FLOATING TANKS
MR. TANIKAZE, HAVE YOU EVER HAD AN ANGLERFISH HOT POT?
I DON'T THINK SO...
IT'S DELICIOUS! I'LL MAKE ONE FOR YOU, SO PLEASE HAVE SOME!
UM, SURE ...
6
I CAN TAKE CARE OF THE BOARDING PROCEDURES AFTER VALIDATION.
DOOR OPENING ...
6
I SEE, IT'S TRANS-PARENT ON THE INSIDE.
TWO PERSONS
BEEP

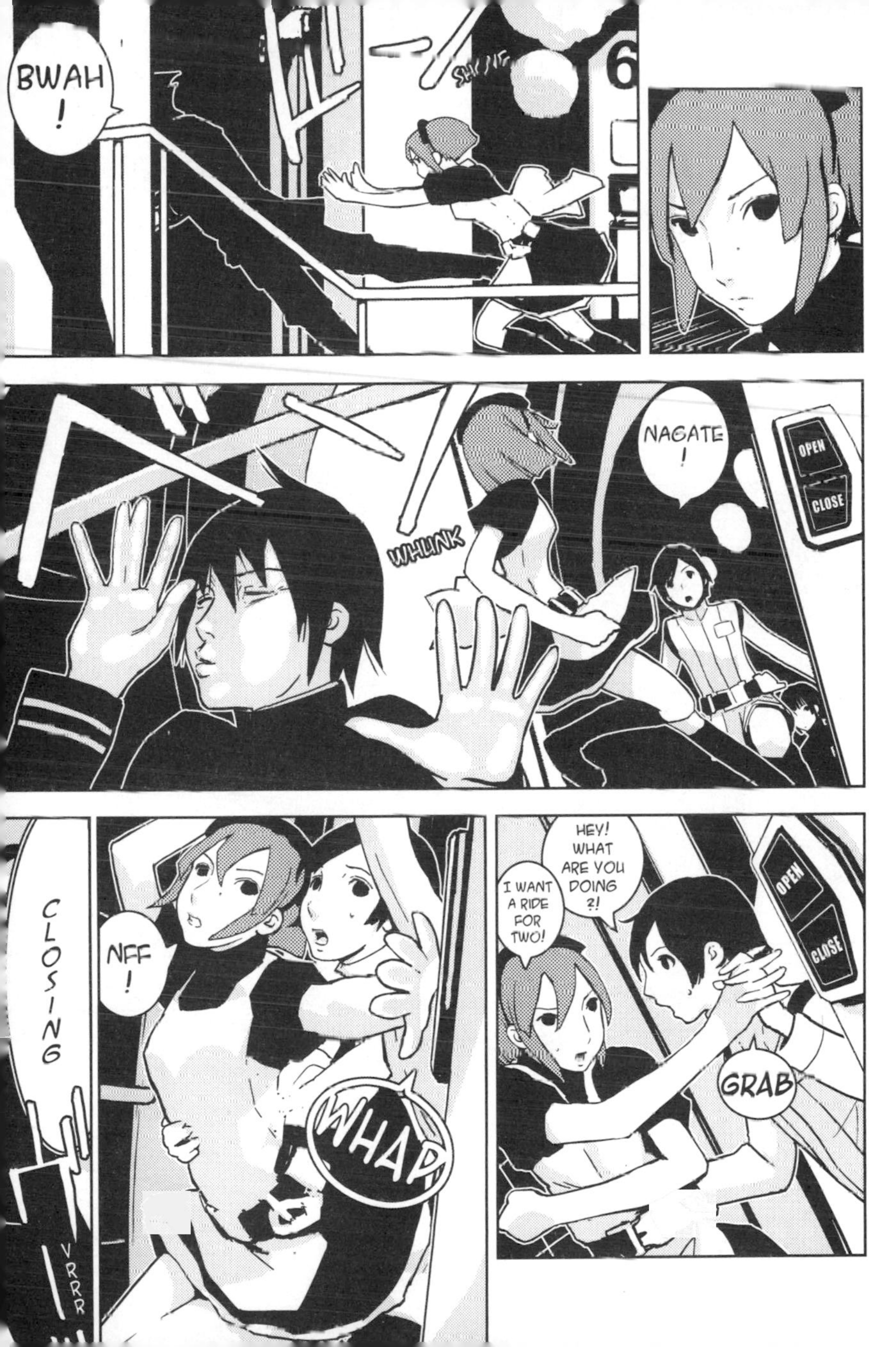
6
BWAH!
NAGATE!
WHUNK
OPEN
CLOSE
HEY! WHAT ARE YOU DOING?!
I WANT A RIDE FOR TWO!
OPEN
CLOSE
GRAB
NFF!
WHAP
CLOSING
VRRR

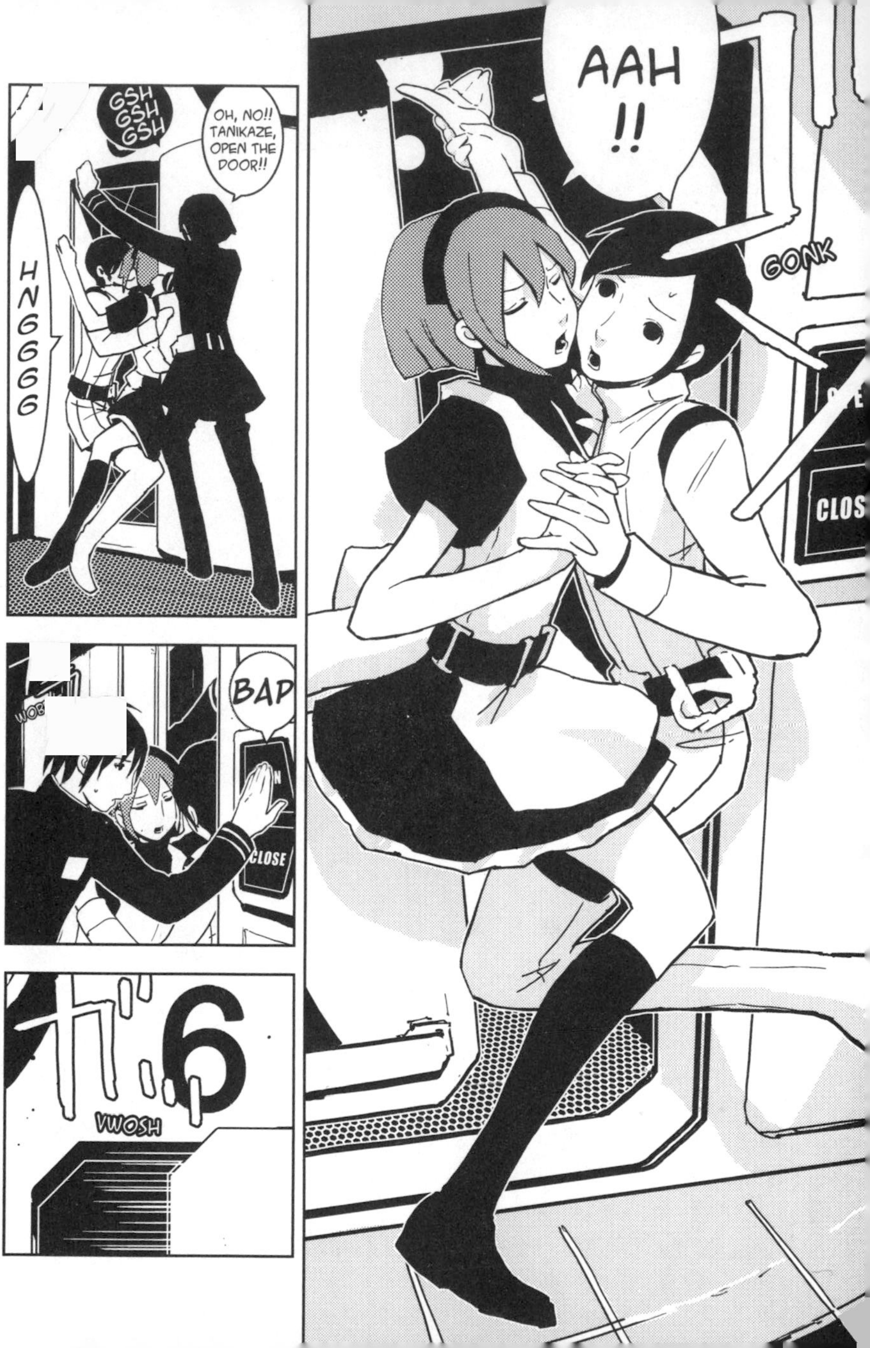
AAH !!
GONK
GSH GSH GSH
OH, NO!! TANIKAZE, OPEN THE DOOR!!
HNGGGG
BAP
CLOSE
VWOSH

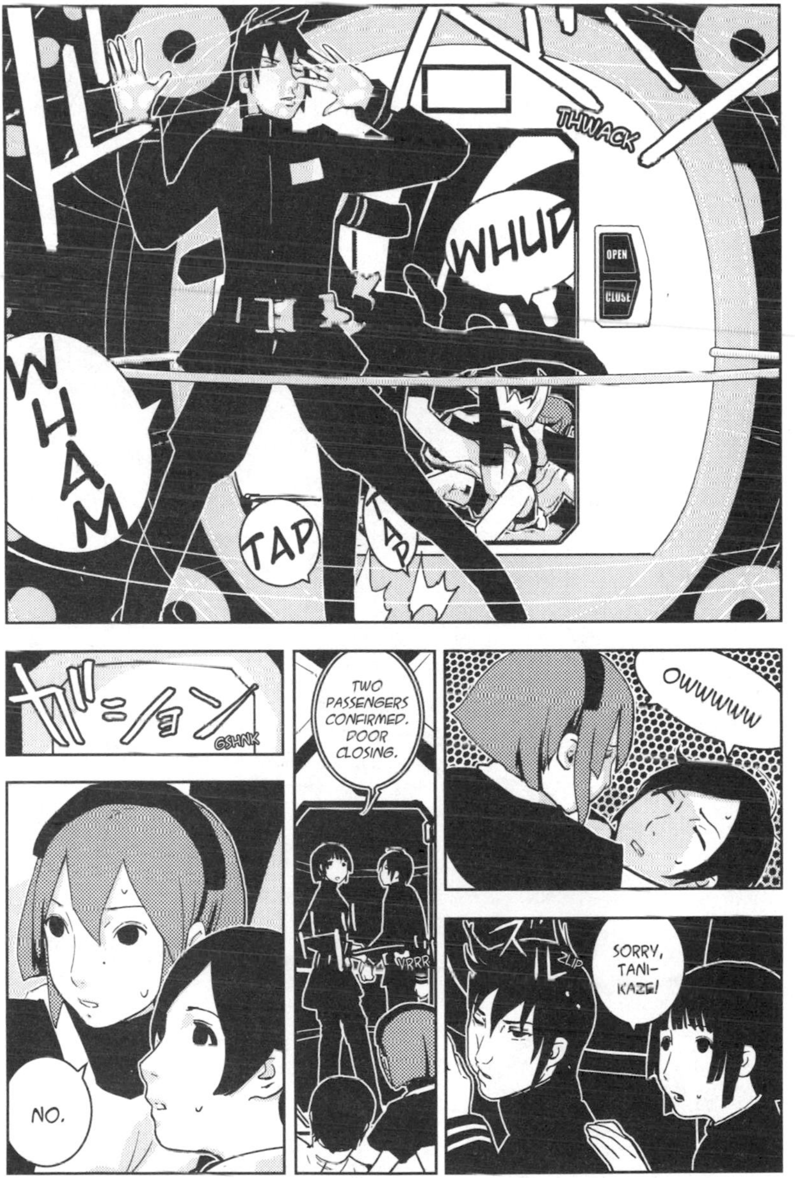
THWACK
WHUD
OPEN
CLOSE
WHAM
TAP
TAP
OWWWWW
ZLIP
SORRY, TANI-KAZE!
TWO PASSENGERS CONFIRMED. DOOR CLOSING.
VRRR
GSHNK
NO.

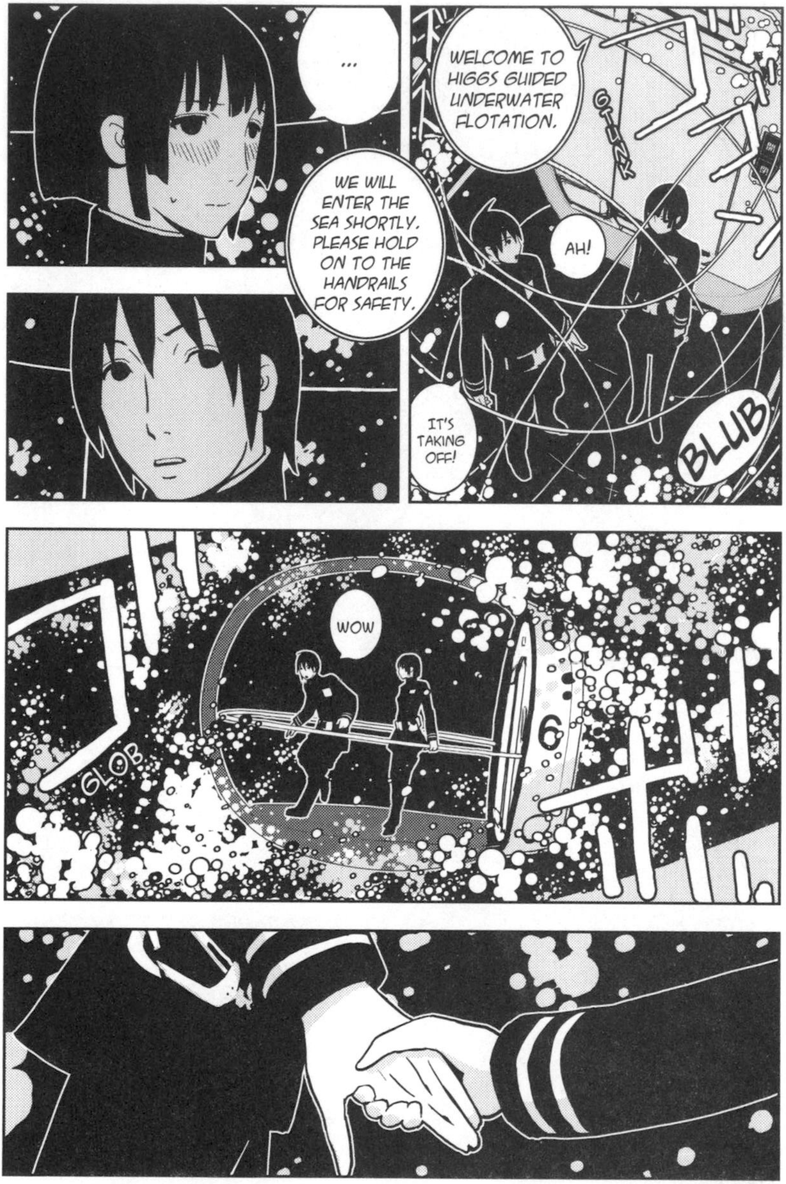
WELCOME TO HIGGS GUIDED UNDERWATER FLOTATION.
WE WILL ENTER THE SEA SHORTLY. PLEASE HOLD ON TO THE HANDRAILS FOR SAFETY.
AH!
IT'S TAKING OFF!
BLUB
...
WOW
GLOB

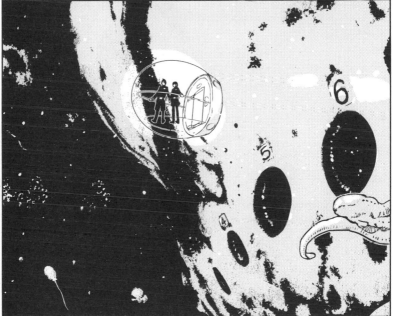

Chapter 8: END

KNIGHTS OF SIDONIA
BY TSUTOMU NIHEI

Chapter 9: Shizuka Hoshijiro's Salute

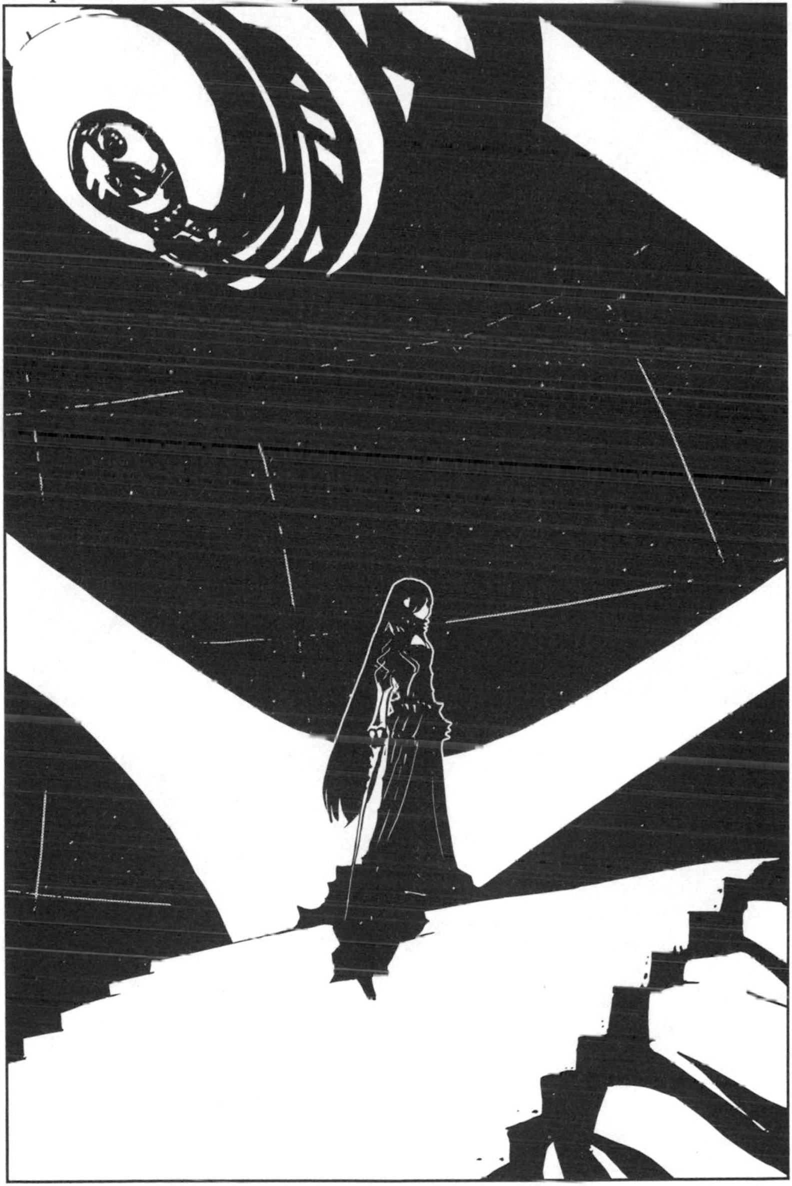

One Hundred Sights of Sidonia Part Seven: The Captain's Chamber

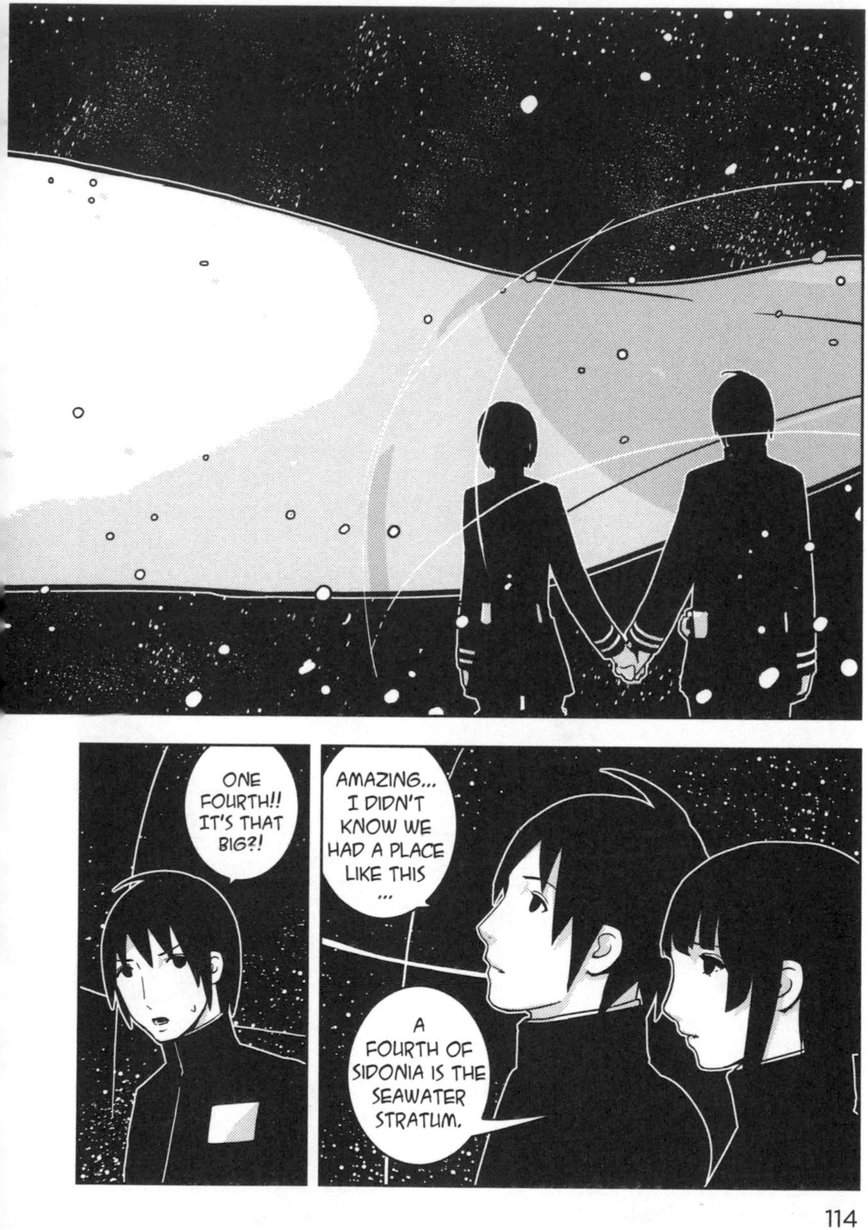
AMAZING... I DIDN'T KNOW WE HAD A PLACE LIKE THIS ...
A FOURTH OF SIDONIA IS THE SEAWATER STRATUM.
ONE FOURTH!! IT'S THAT BIG?!

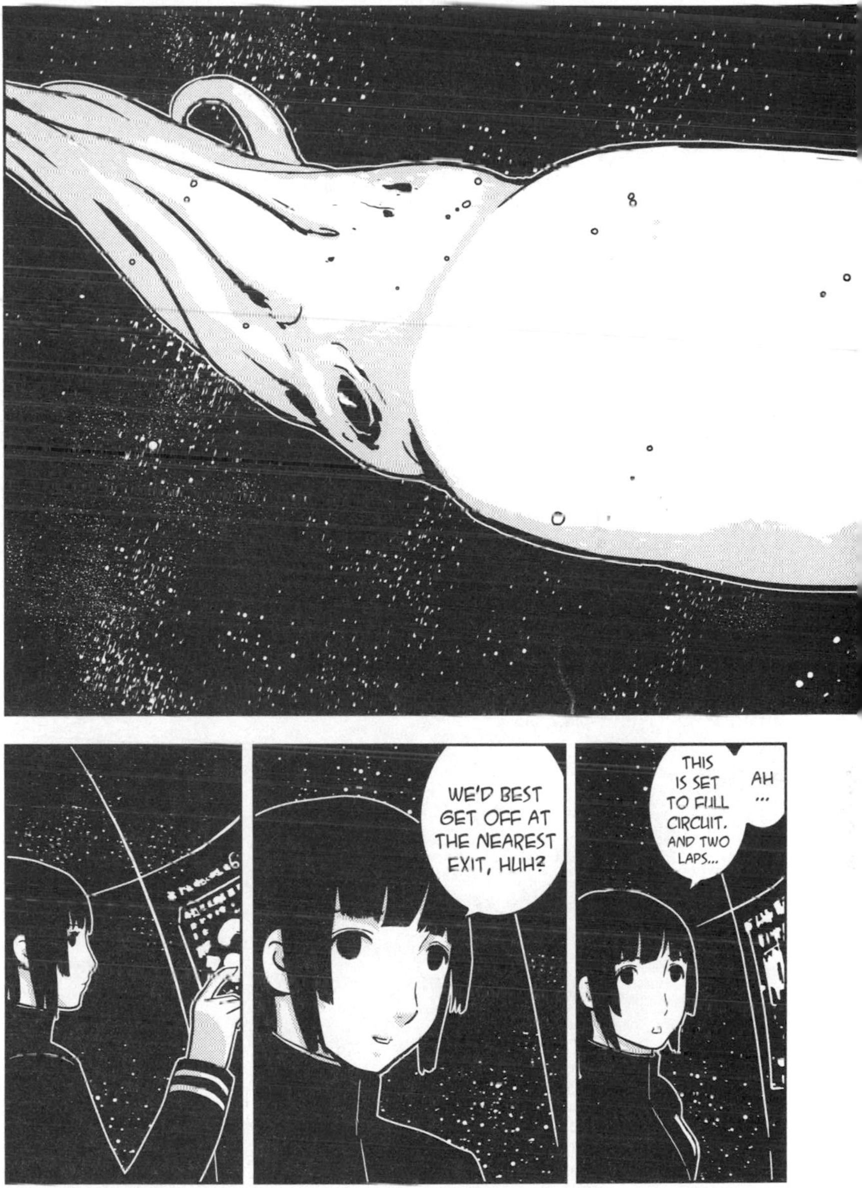
AH ...
THIS IS SET TO FULL CIRCUIT. AND TWO LAPS...
WE'D BEST GET OFF AT THE NEAREST EXIT, HUH?

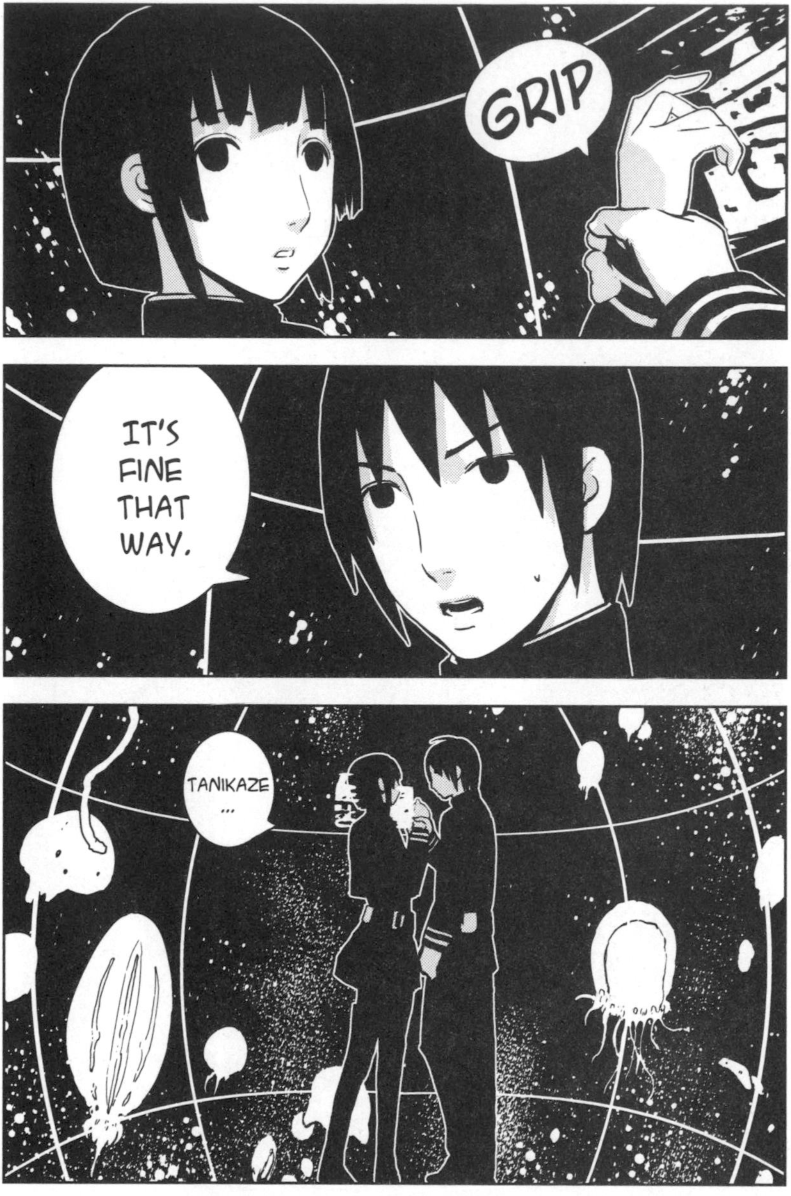
GRIP
IT'S FINE THAT WAY.
TANIKAZE ...

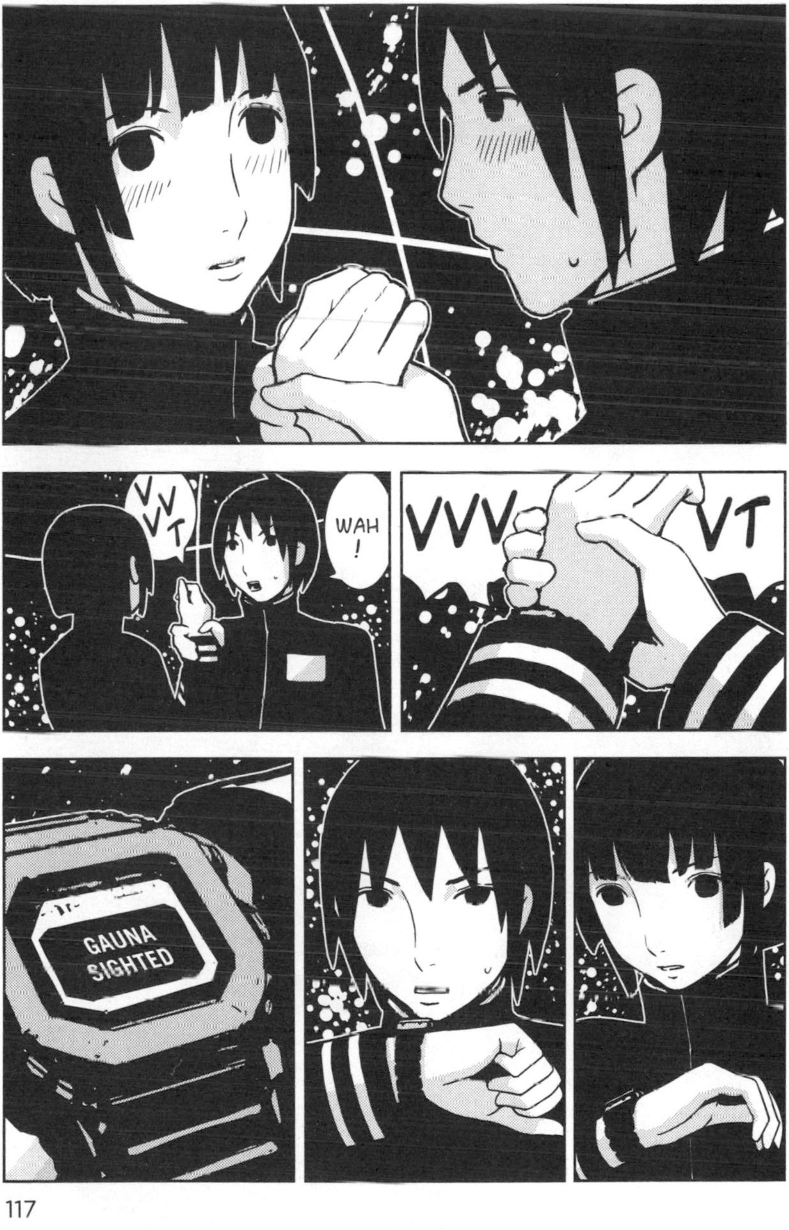
VVV VT
WAH!
VVVT
GAUNA SIGHTED

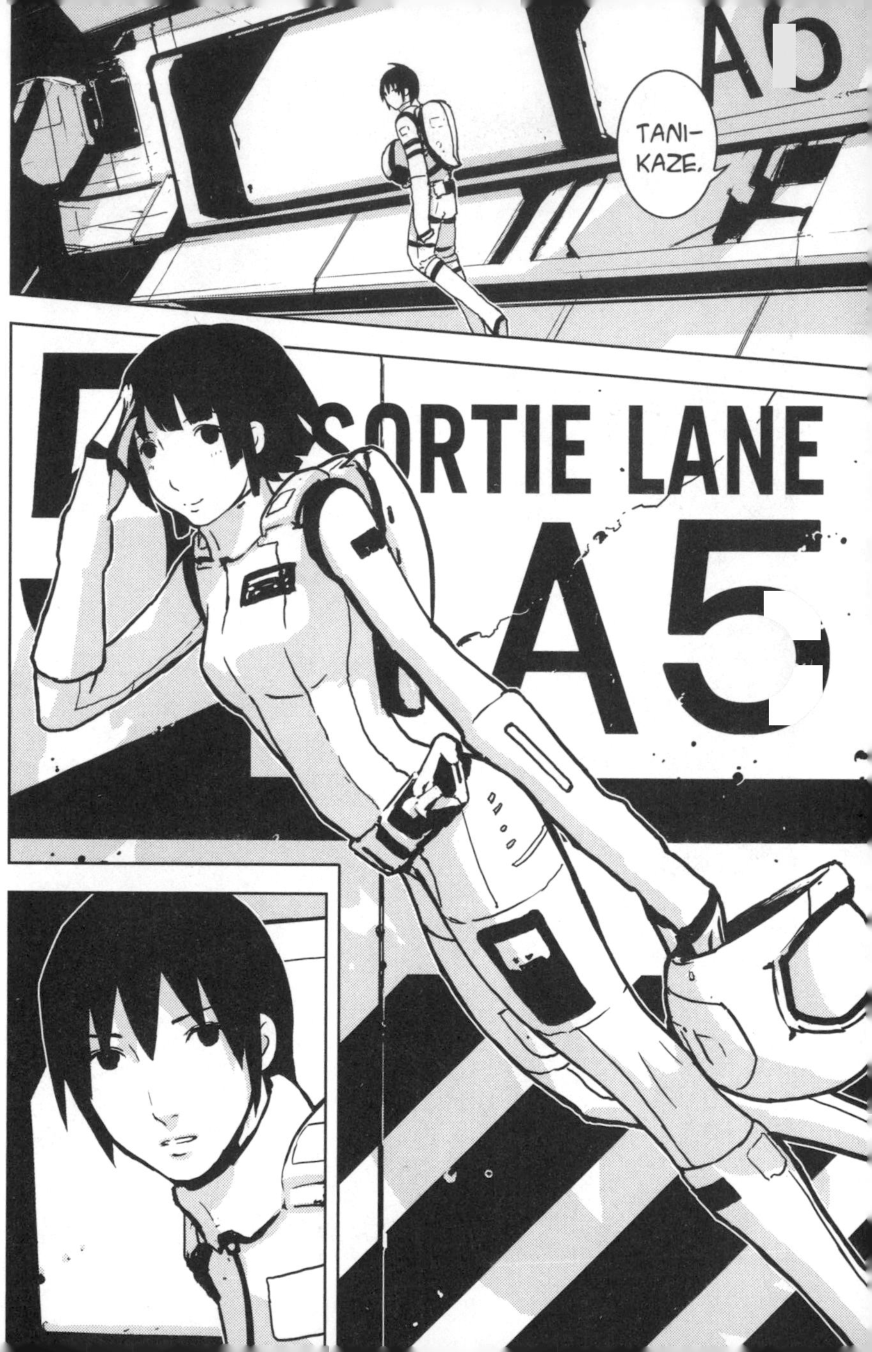
TANI-KAZE.
SORTIE LANE
A5

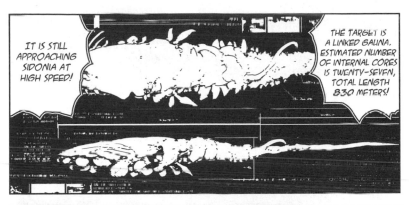

THE TARGET IS A LINKED GAUNA. ESTIMATED NUMBER OF INTERNAL CORES IS TWENTY-SEVEN, TOTAL LENGTH 830 METERS!
IT IS STILL APPROACHING SIDONIA AT HIGH SPEED!
THIRTY-SIX SUPPRESSION FORCE GARDES, EQUIPPED WITH EIGHT KABIZASHIS FOR A FORCE RATIO OF 3.37!*
G 488
THERE IS ALWAYS A PRIMAL CORE IN A LINKED GAUNA! WHEN YOU FIND IT, DESTROY IT AS YOUR UTMOST PRIORITY!
PRIMAL CORE ESTIMATED POSITION
FOLLOW-UP INFORMATION FROM THE SCOUT UNIT. IT IS ALMOST CERTAIN THAT THE REAR 250 METERS OF THE TARGET DO NOT CONTAIN ANY CORES.
CUT IT OFF, AND WE MASSIVELY DIMINISH THE VOLUME OF PLACENTA. KUNATO SQUAD, SEVER THE TAIL AS PLANNED!!
701
ROGER!
702
ROGER!
703
ROGER!

* the ratio of 27 Gaunas to 8 Kabizashis

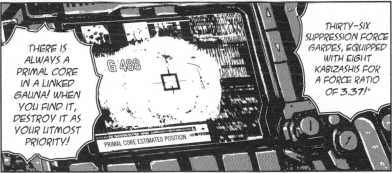

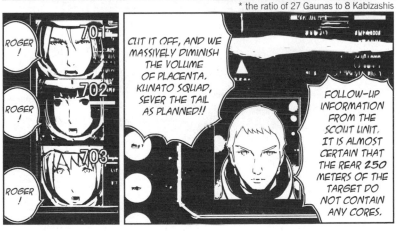

ROGER
!

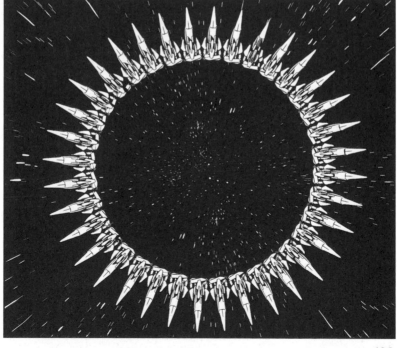

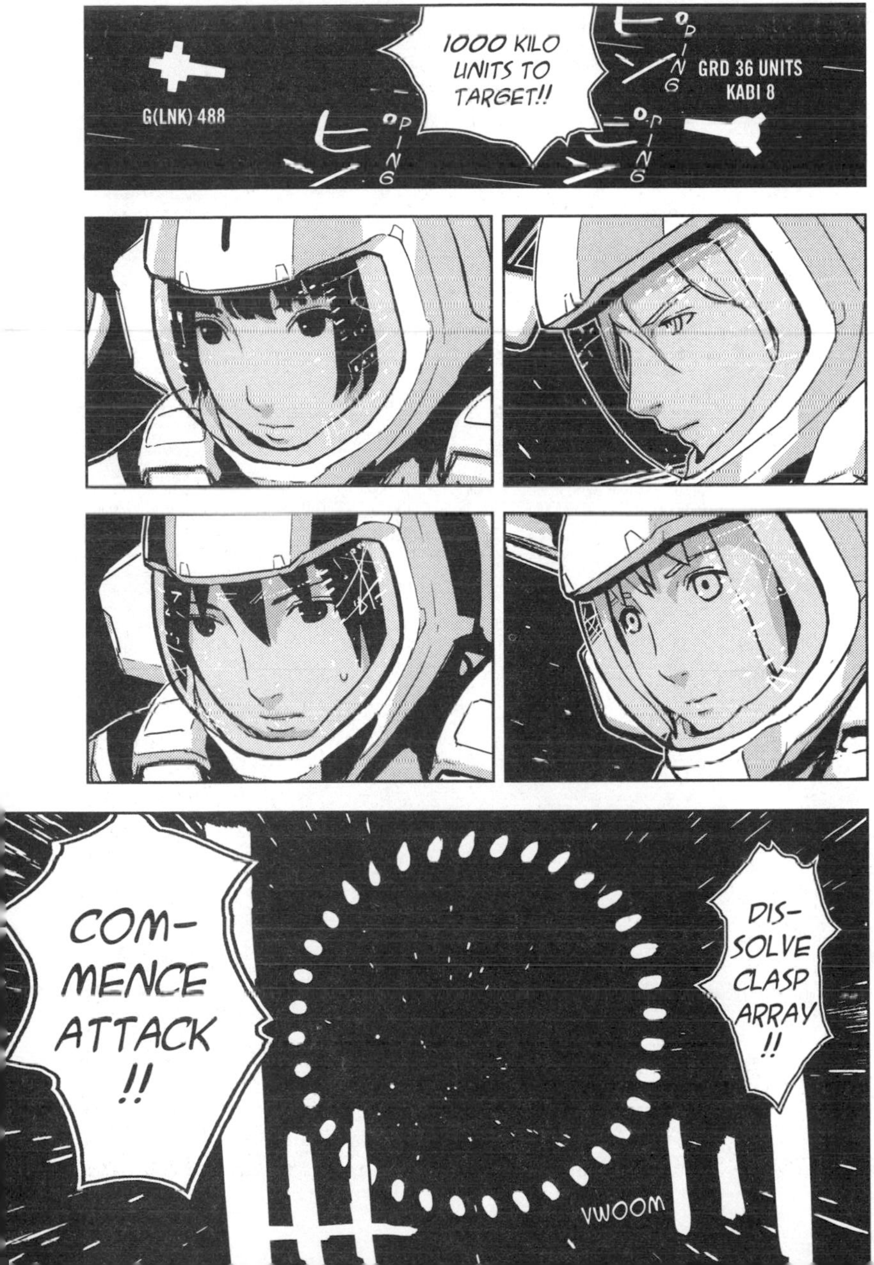
PING
GRD 36 UNITS
KABI 8
1000 KILO UNITS TO TARGET!!
PING
PING
G(LNK) 488
DIS-SOLVE CLASP ARRAY!!
COM-MENCE ATTACK!!
VWOOM

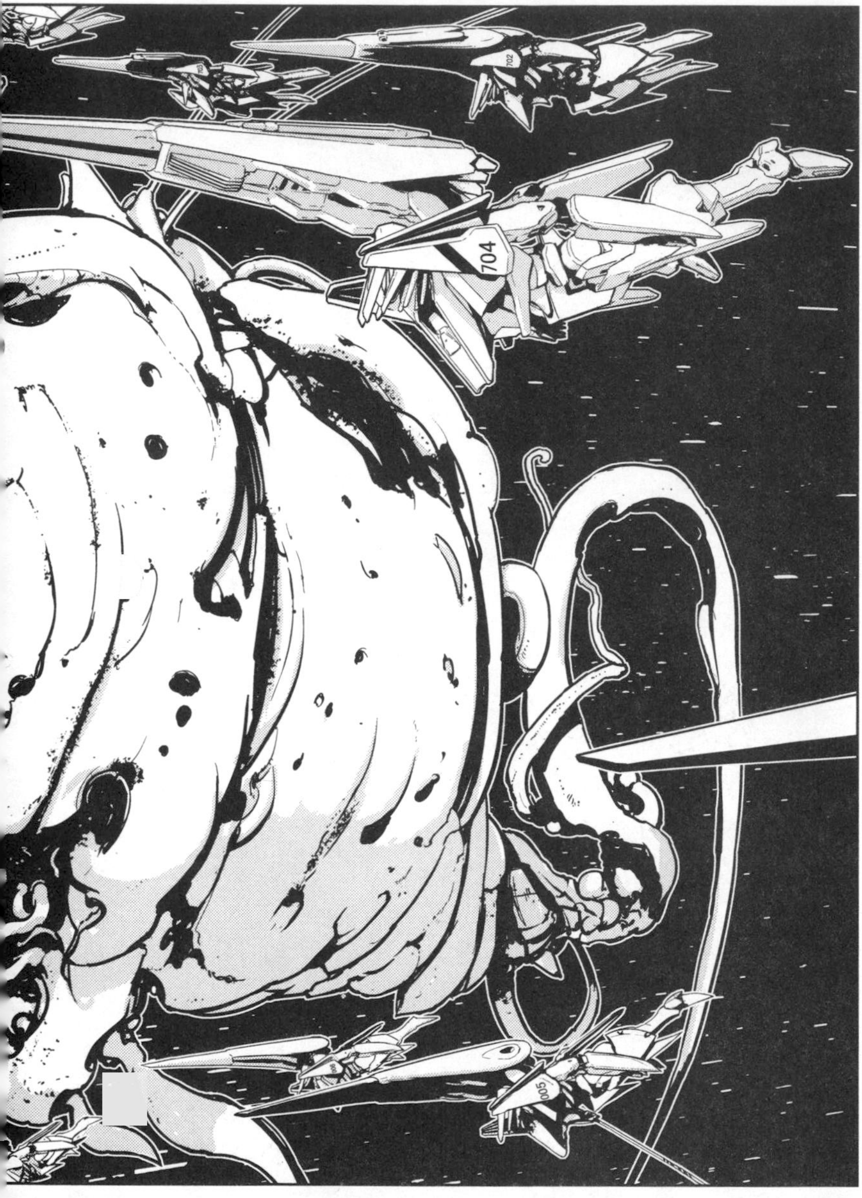
704

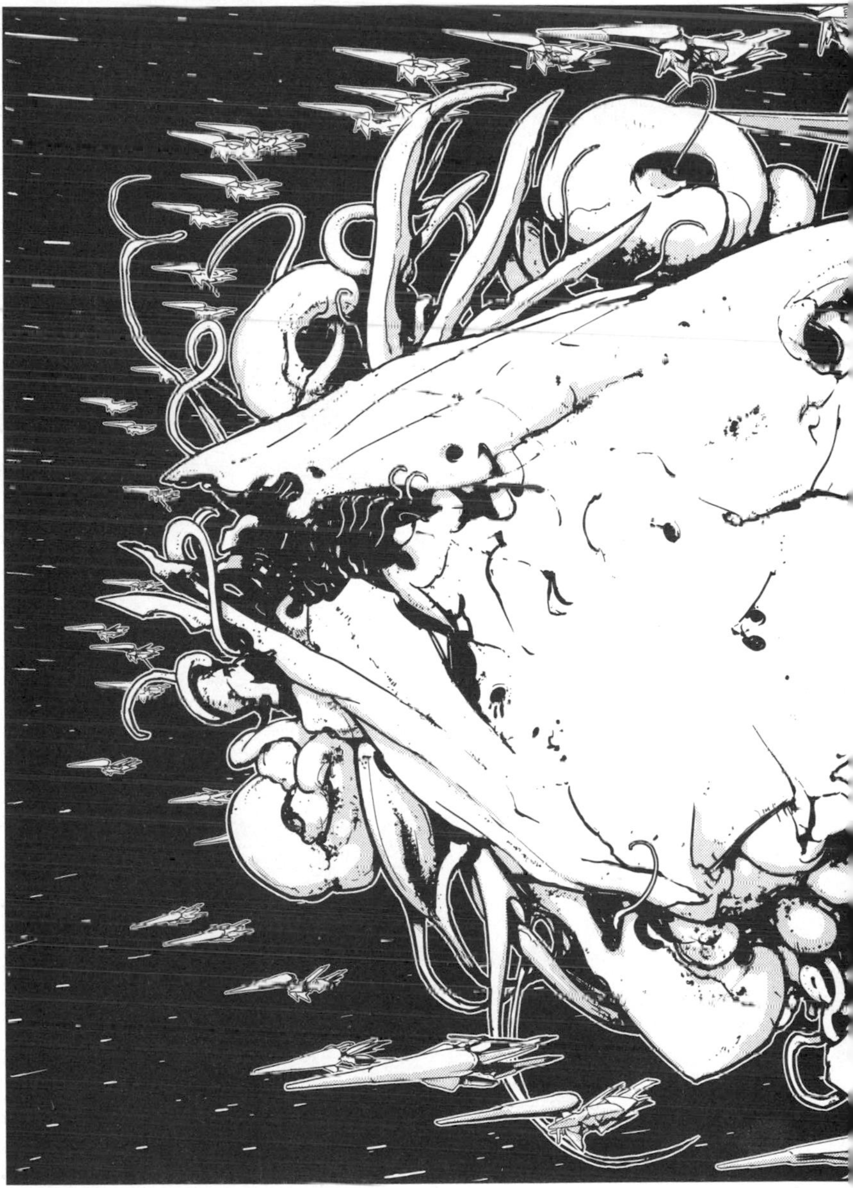

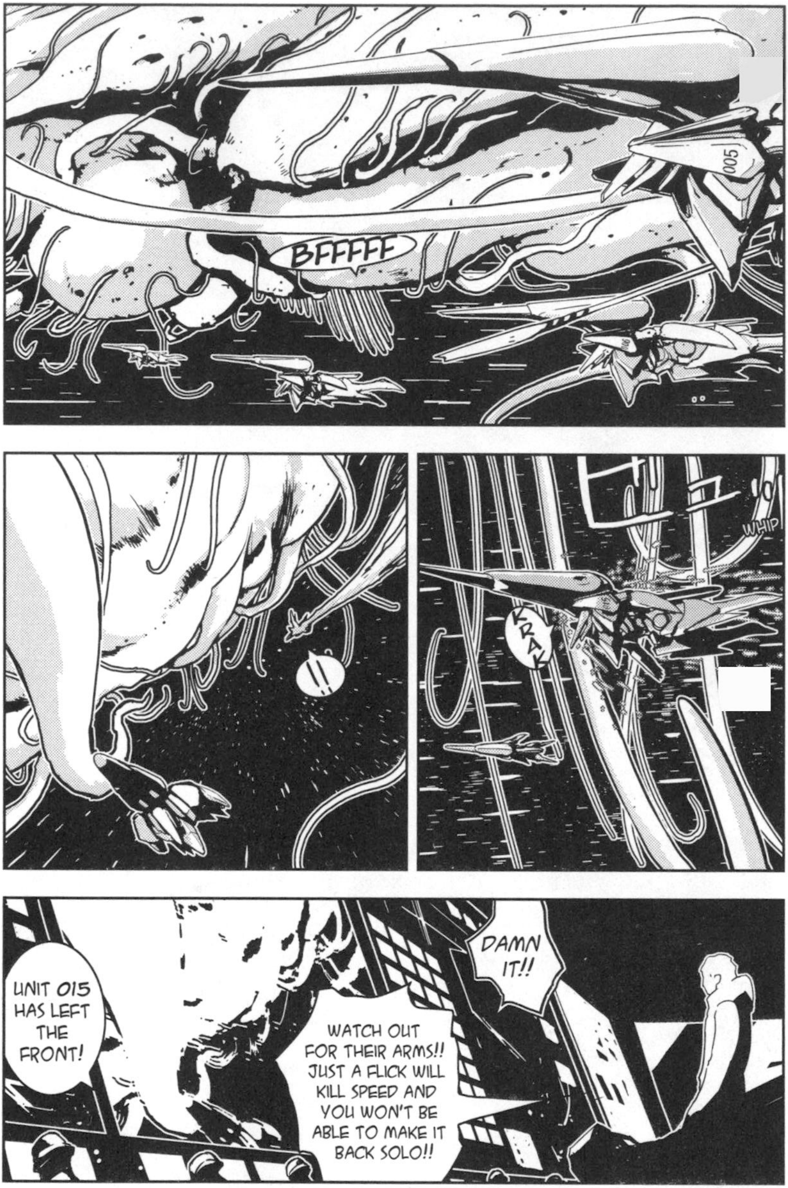
005
BFFFFF
WHIP
KRAK
!!
DAMN IT!!
WATCH OUT FOR THEIR ARMS!! JUST A FLICK WILL KILL SPEED AND YOU WON'T BE ABLE TO MAKE IT BACK SOLO!!
UNIT 015 HAS LEFT THE FRONT!

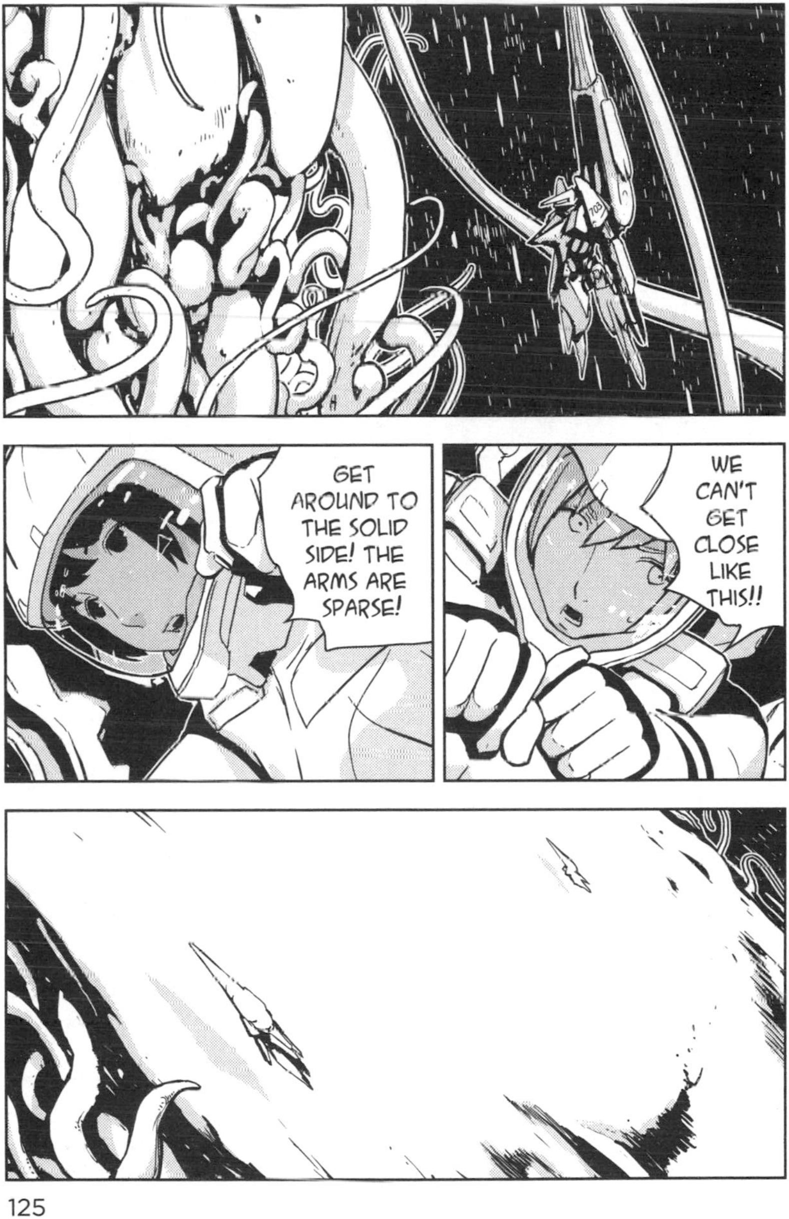
WE CAN'T GET CLOSE LIKE THIS!!
GET AROUND TO THE SOLID SIDE! THE ARMS ARE SPARSE!

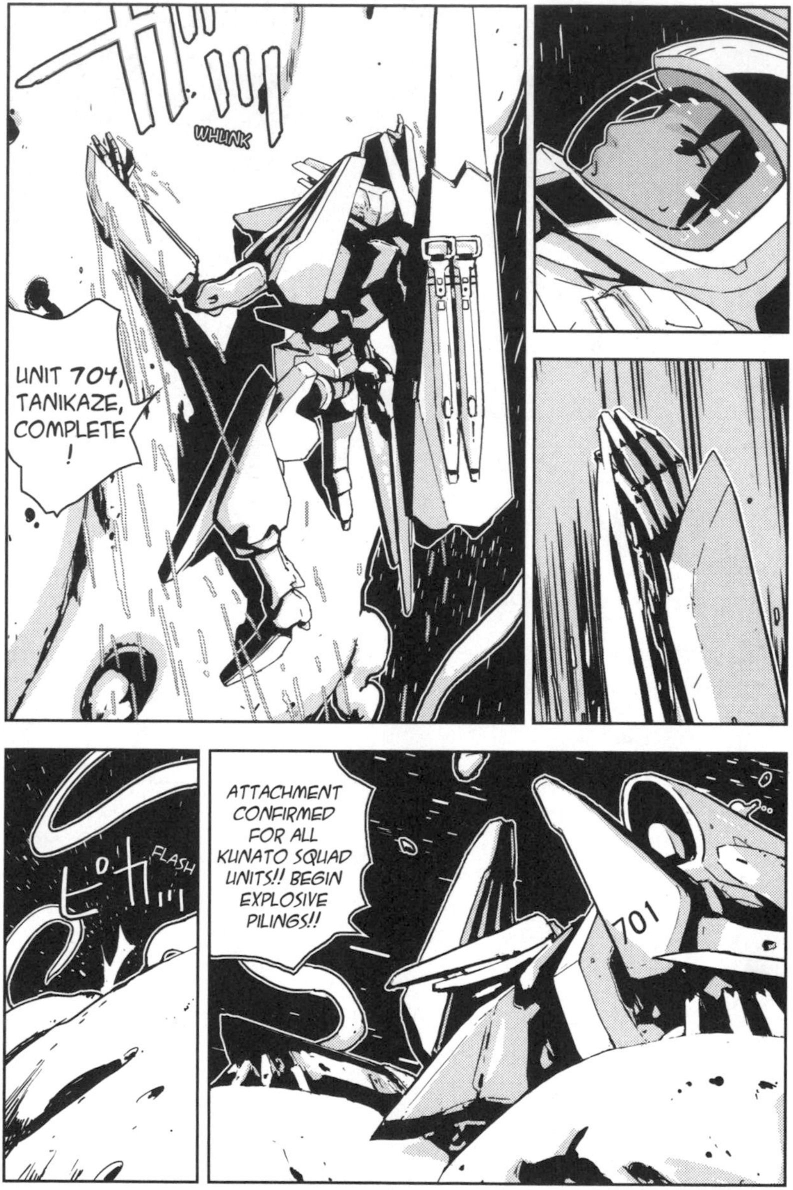
WHUNK
UNIT 704, TANIKAZE, COMPLETE!
ATTACHMENT CONFIRMED FOR ALL KUNATO SQUAD UNITS!! BEGIN EXPLOSIVE PILINGS!!
701
FLASH

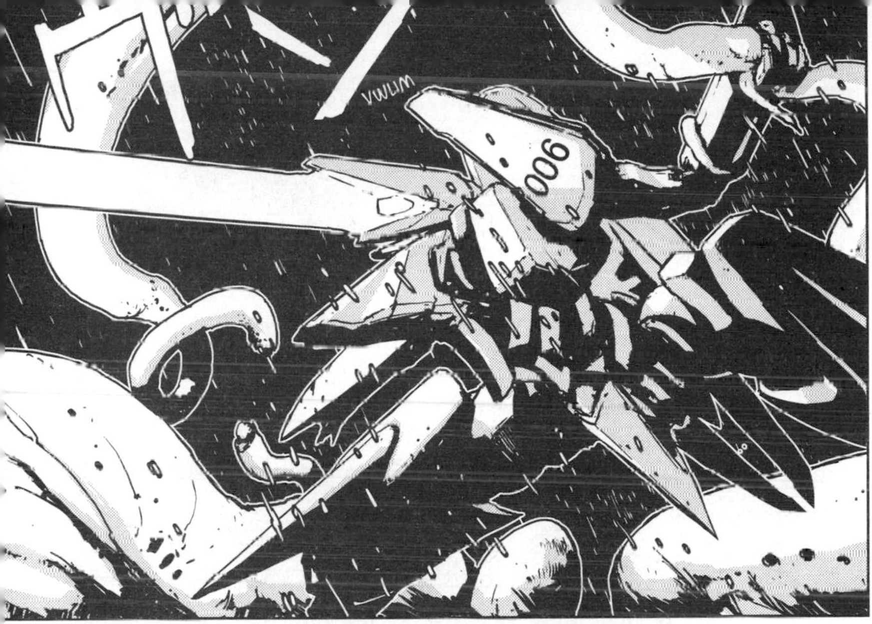
VWLIM
006
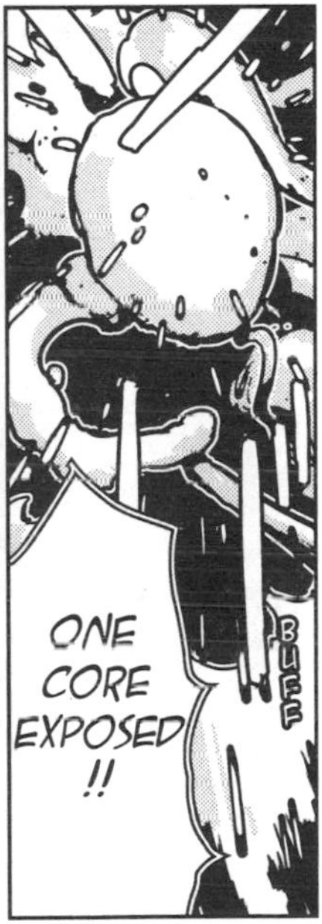
BUFF
ONE CORE EXPOSED !!
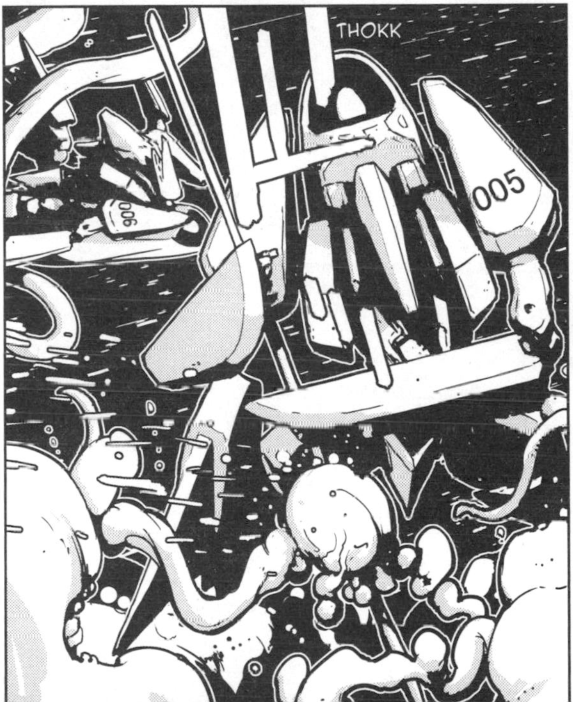
THOKK
005
006

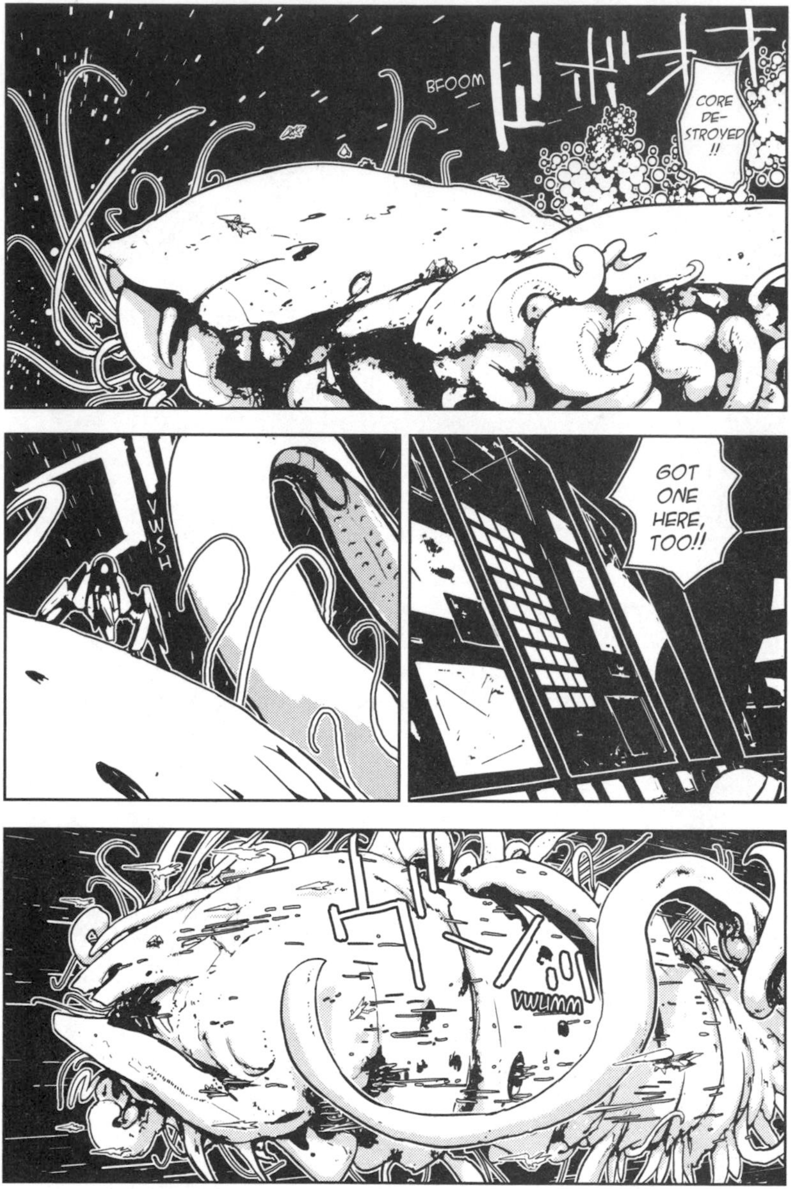
CORE DE-STROYED!!
ドボオオ
BFOOM
GOT ONE HERE, TOO!!
VWSH
VWUMM

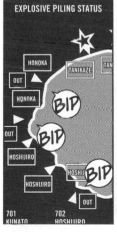

EXPLOSIVE PILING STATUS

HONOKA

OUT

HONOKA

TANIKAZE

TAN

OUT

HOSHIJIRO

HOSHIJIRO

HOSHIJI

OUT

BIP

BIP

BIP

701
KUNATO

702
HOSHIJIRO

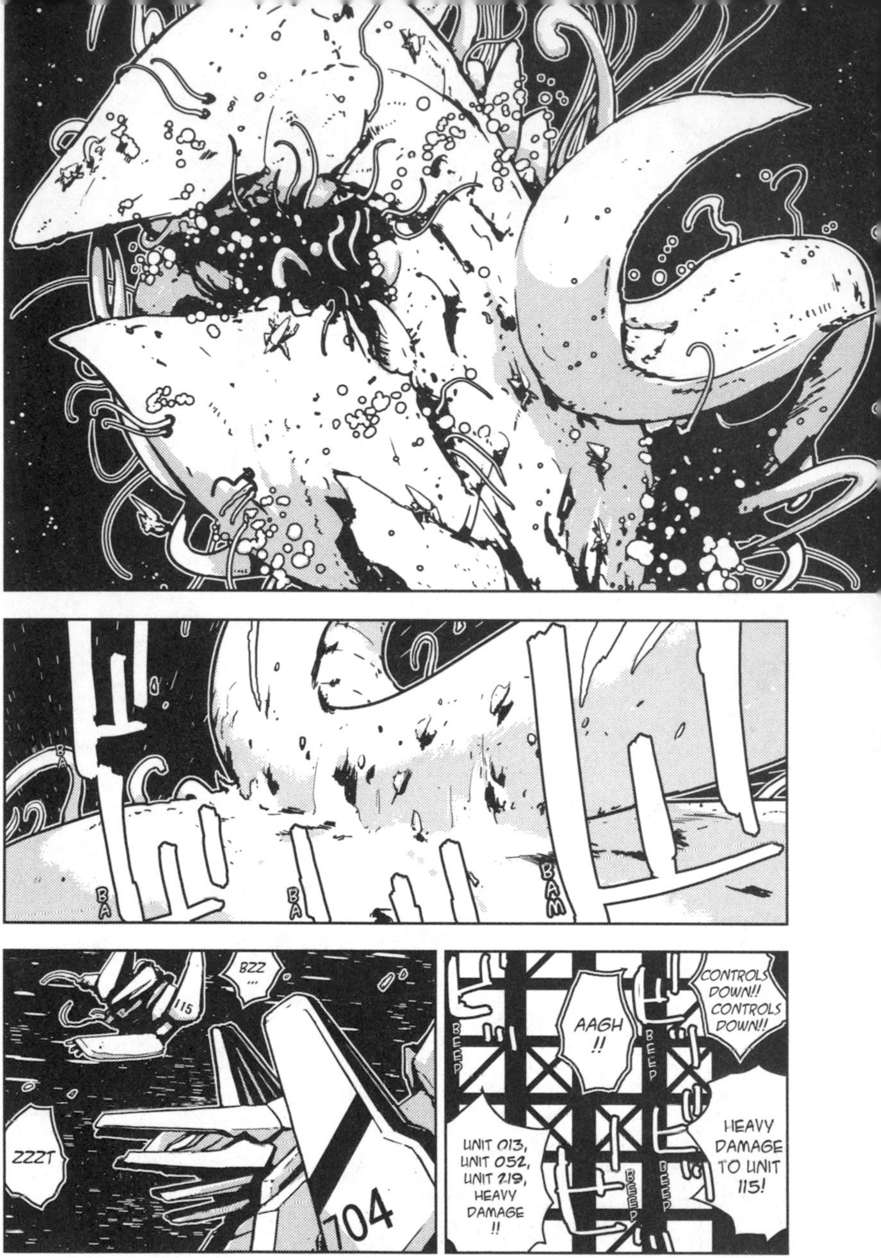
BAM
BA
BA
BA
CONTROLS DOWN!! CONTROLS DOWN!!
BEEP
AAGH !!
BEEP
HEAVY DAMAGE TO UNIT 115!
BEEP
BEEP
UNIT 013, UNIT 052, UNIT 219, HEAVY DAMAGE !!
BEEP
BZZ ...
115
ZZZT
104

TANI-KAZE!!

W-WAIT...

NO...
I...

WHOK

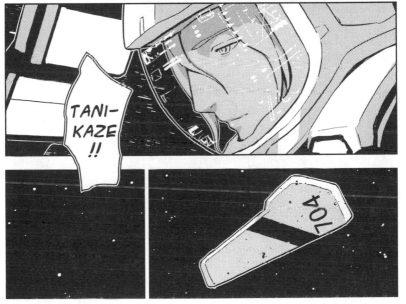

TANI-KAZE!!

704

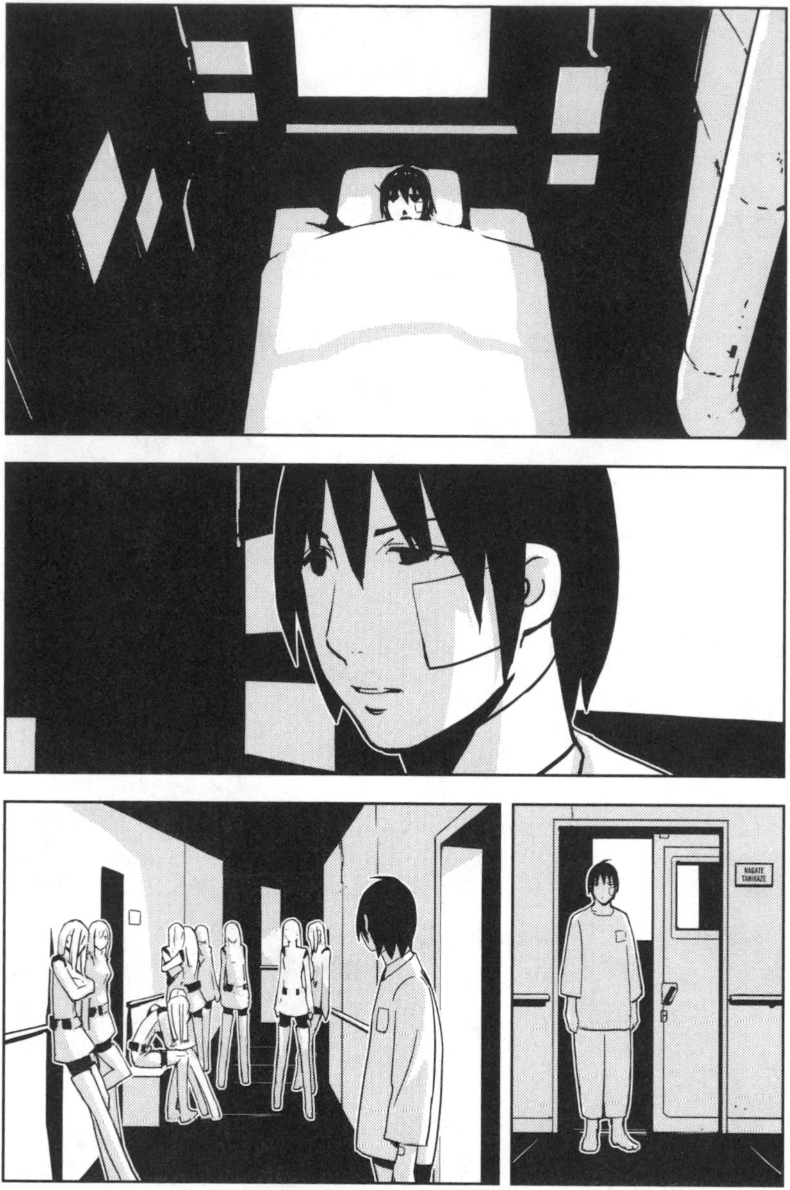
NAGATE
TANIKAZE

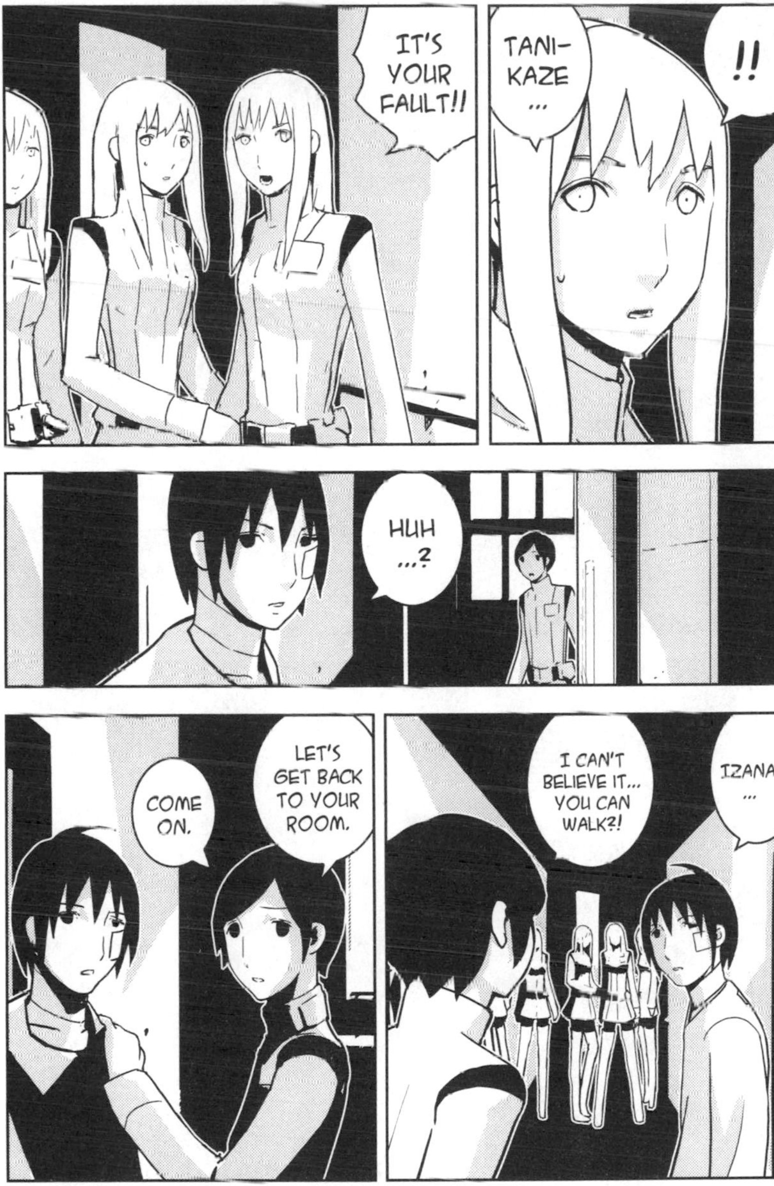
!!
TANI-KAZE ...
IT'S YOUR FAULT!!
HUH ...?
IZANA ...
I CAN'T BELIEVE IT... YOU CAN WALK?!
LET'S GET BACK TO YOUR ROOM.
COME ON.

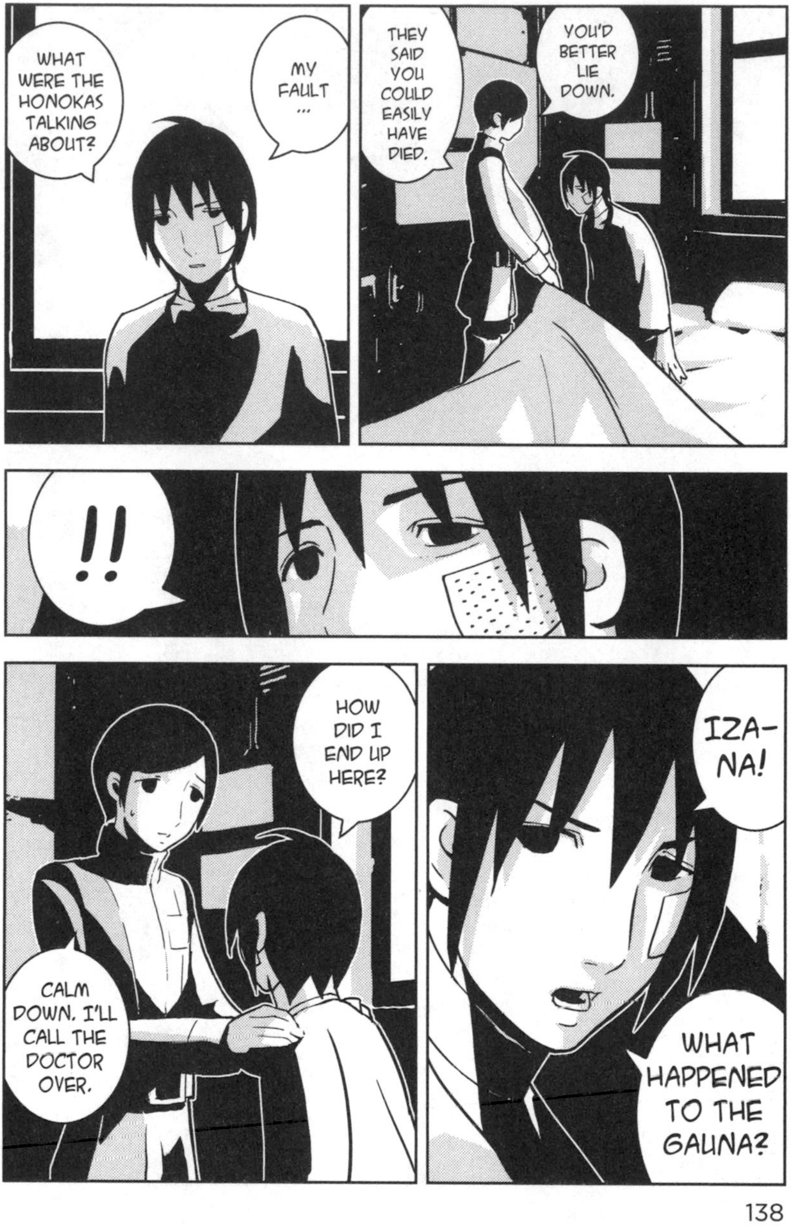
YOU'D BETTER LIE DOWN.
THEY SAID YOU COULD EASILY HAVE DIED.
MY FAULT ...
WHAT WERE THE HONOKAS TALKING ABOUT?
!!
IZA-NA!
WHAT HAPPENED TO THE GAUNA?
HOW DID I END UP HERE?
CALM DOWN. I'LL CALL THE DOCTOR OVER.

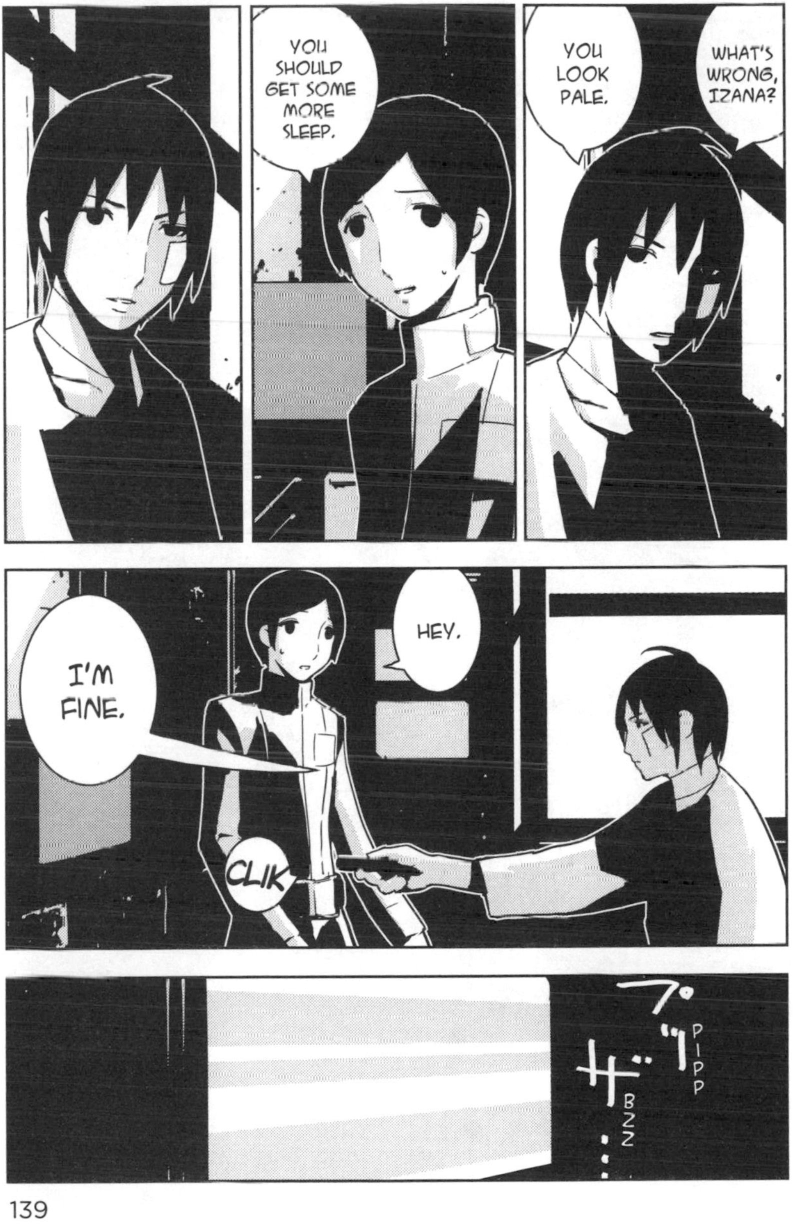
WHAT'S WRONG, IZANA?
YOU LOOK PALE.
YOU SHOULD GET SOME MORE SLEEP.
HEY.
I'M FINE.
CLIK
プツ PIPP
ザ BZZ ...

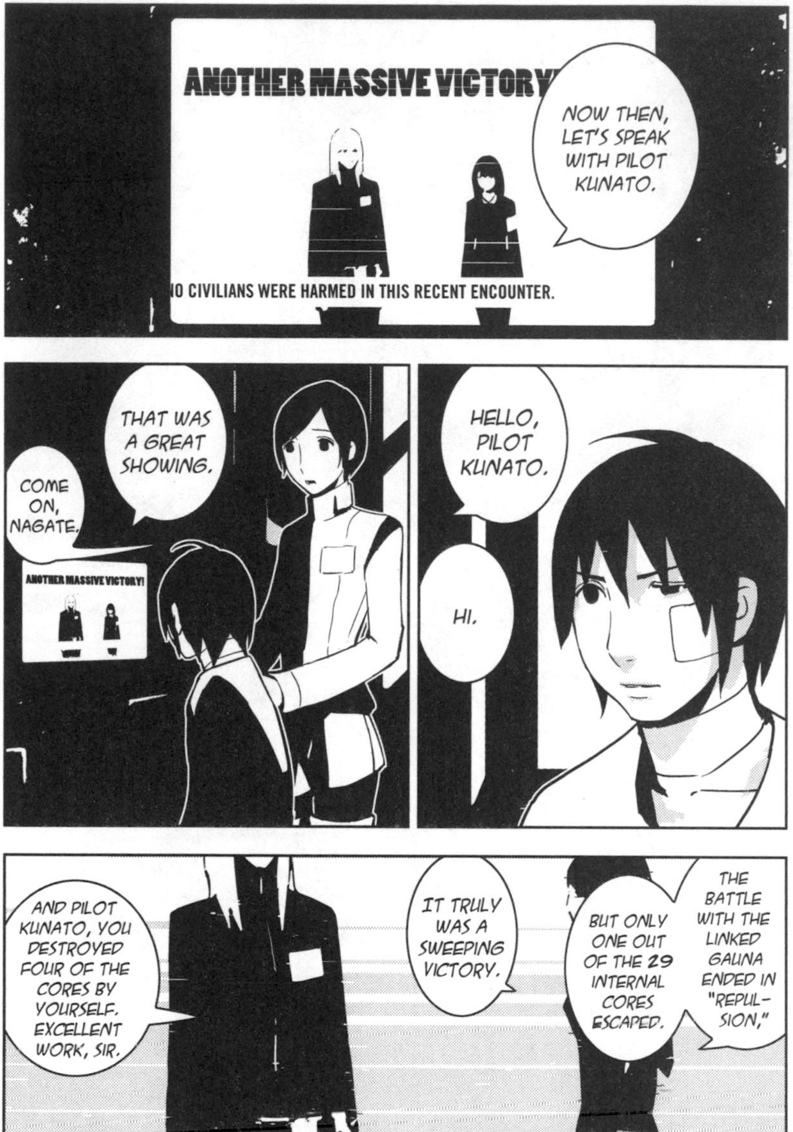

ANOTHER MASSIVE VICTORY
NOW THEN, LET'S SPEAK WITH PILOT KUNATO.
NO CIVILIANS WERE HARMED IN THIS RECENT ENCOUNTER.
HELLO, PILOT KUNATO.
HI.
THAT WAS A GREAT SHOWING.
COME ON, NAGATE.
ANOTHER MASSIVE VICTORY!
THE BATTLE WITH THE LINKED GAUNA ENDED IN "REPULSION,"
BUT ONLY ONE OUT OF THE 29 INTERNAL CORES ESCAPED.
IT TRULY WAS A SWEEPING VICTORY.
AND PILOT KUNATO, YOU DESTROYED FOUR OF THE CORES BY YOURSELF. EXCELLENT WORK, SIR.
EN HONOKA: CRITICAL, UNCONSCIOUS
NAGATE TANIKAZE: HIGHLY CRITICAL, UNCONSCIOUS

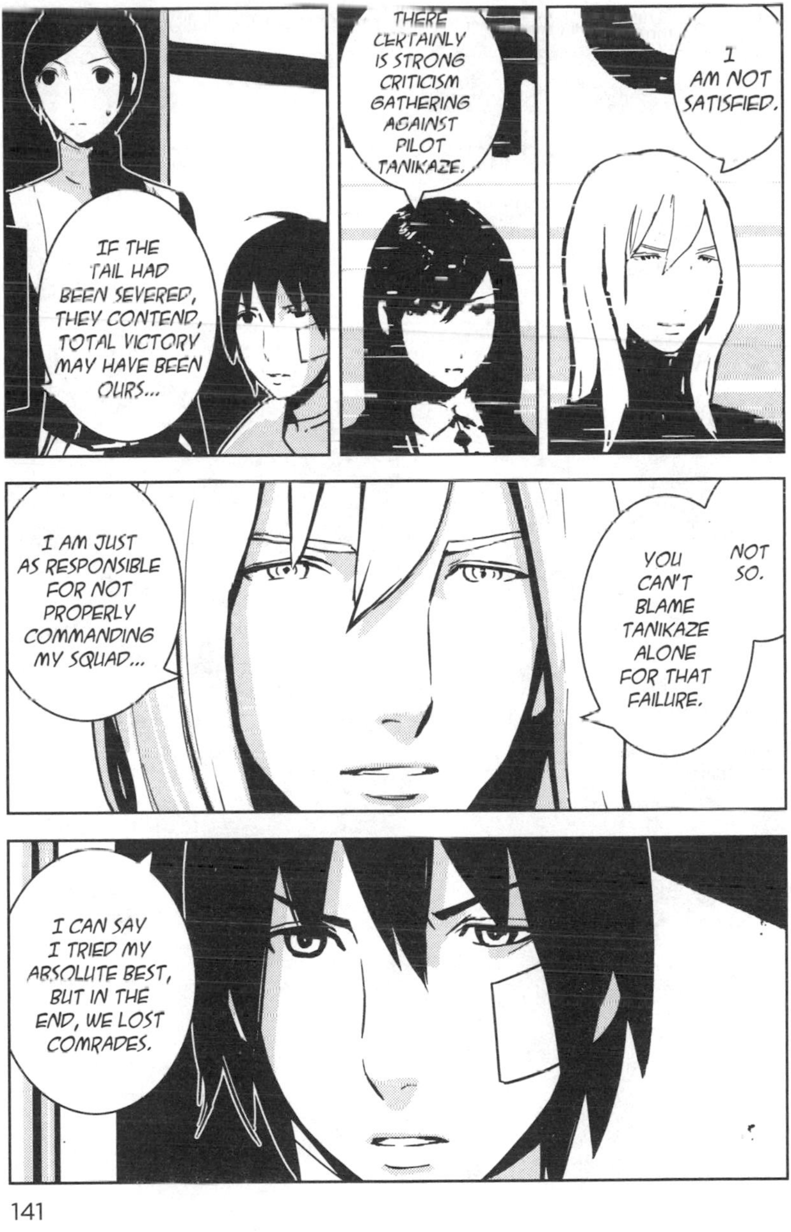
I AM NOT SATISFIED.
THERE CERTAINLY IS STRONG CRITICISM GATHERING AGAINST PILOT TANIKAZE.
IF THE TAIL HAD BEEN SEVERED, THEY CONTEND, TOTAL VICTORY MAY HAVE BEEN OURS...
NOT SO.
YOU CAN'T BLAME TANIKAZE ALONE FOR THAT FAILURE.
I AM JUST AS RESPONSIBLE FOR NOT PROPERLY COMMANDING MY SQUAD...
I CAN SAY I TRIED MY ABSOLUTE BEST, BUT IN THE END, WE LOST COMRADES.

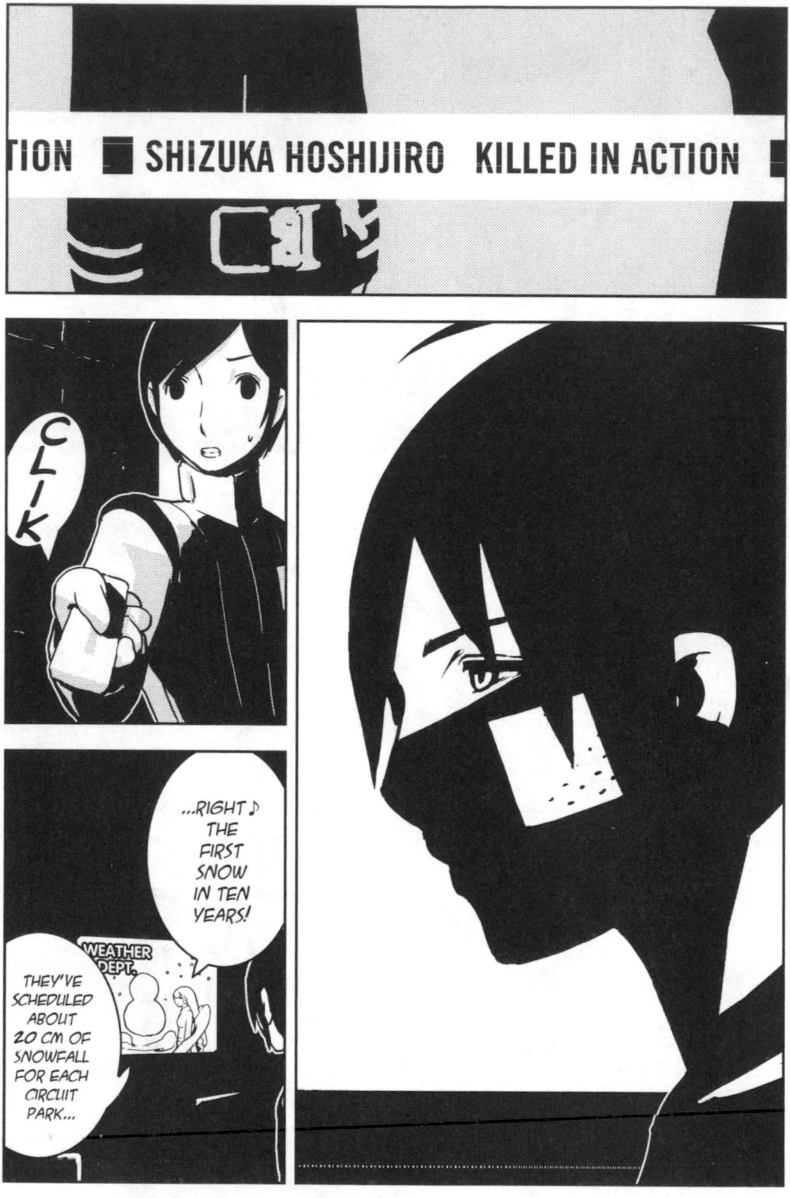
TION ■ SHIZUKA HOSHIJIRO KILLED IN ACTION ■
CLIK
...RIGHT♪ THE FIRST SNOW IN TEN YEARS!
WEATHER DEPT.
THEY'VE SCHEDULED ABOUT 20 CM OF SNOWFALL FOR EACH CIRCUIT PARK...

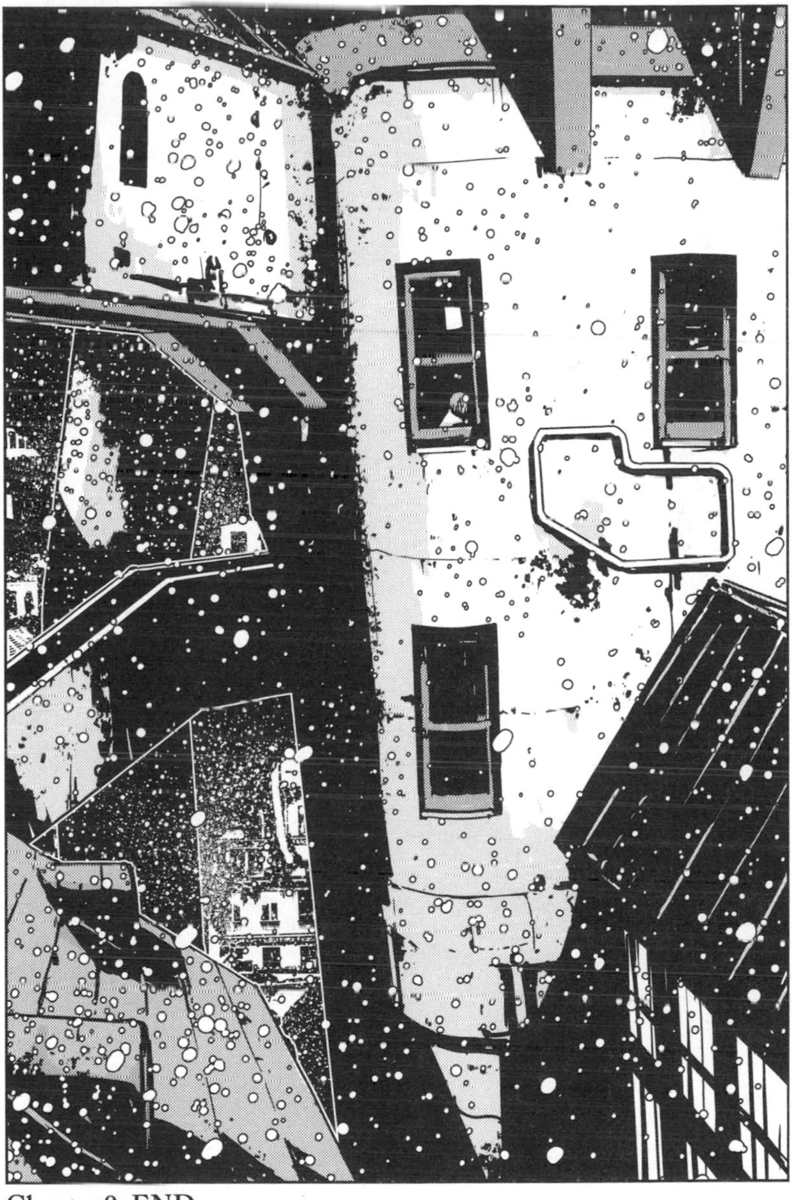

Chapter 9: END

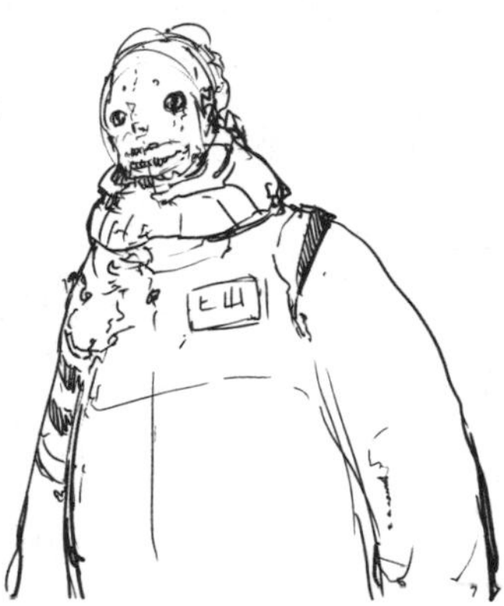
ヒ山

KNIGHTS OF SIDONIA
BY TSUTOMU NIHEI

Chapter 10: The Twenty-Eighth Captain's Decision

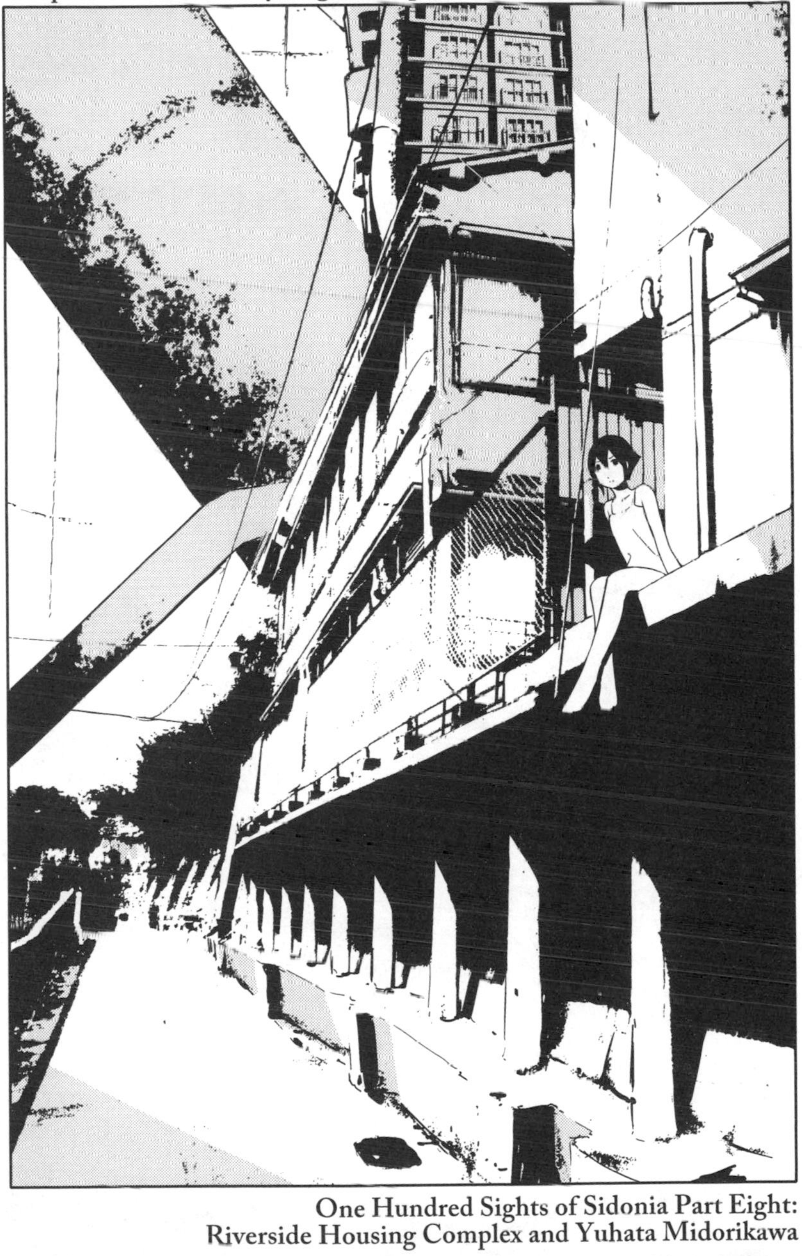

One Hundred Sights of Sidonia Part Eight:
Riverside Housing Complex and Yuhata Midorikawa

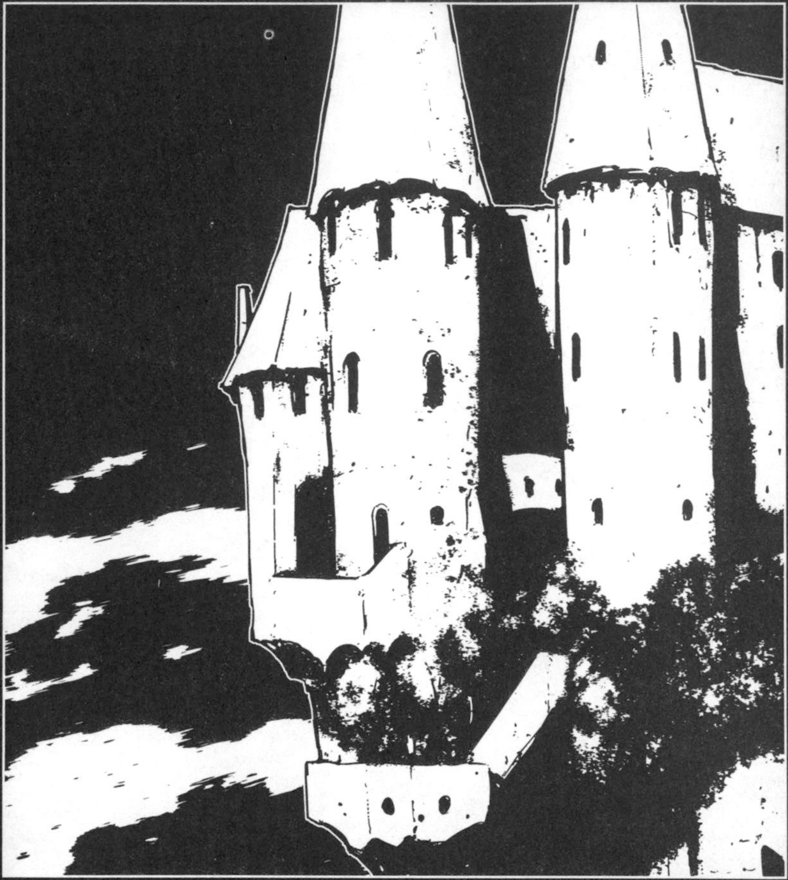

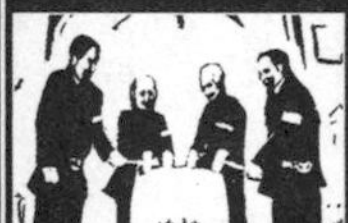

KUNATO DEVELOPMENTS RESEARCH LAB:
RIGHTS ACQUIRED TO PRODUCE NEXT GENERATION GARDES
THE NEXT GENERATION OF GARDES (SERIES 18) IS PLANNED TO SUCCEED THE BASIC FRAMEWORK OF SERIES 17, KNOWN AS THE GREATEST MASTERPIECE IN GARDE HISTORY...
TOHA HEAVY INDUSTRIES: OPERATIONS OF FACTORIES NO. 3 AND NO. 6 TO BE HALTED
TOHA HEAVY INDUSTRIES: MAJORITY OF ENGINEERS IN GARDE DEV DIVISION TO TRANSFER TO KUNATO

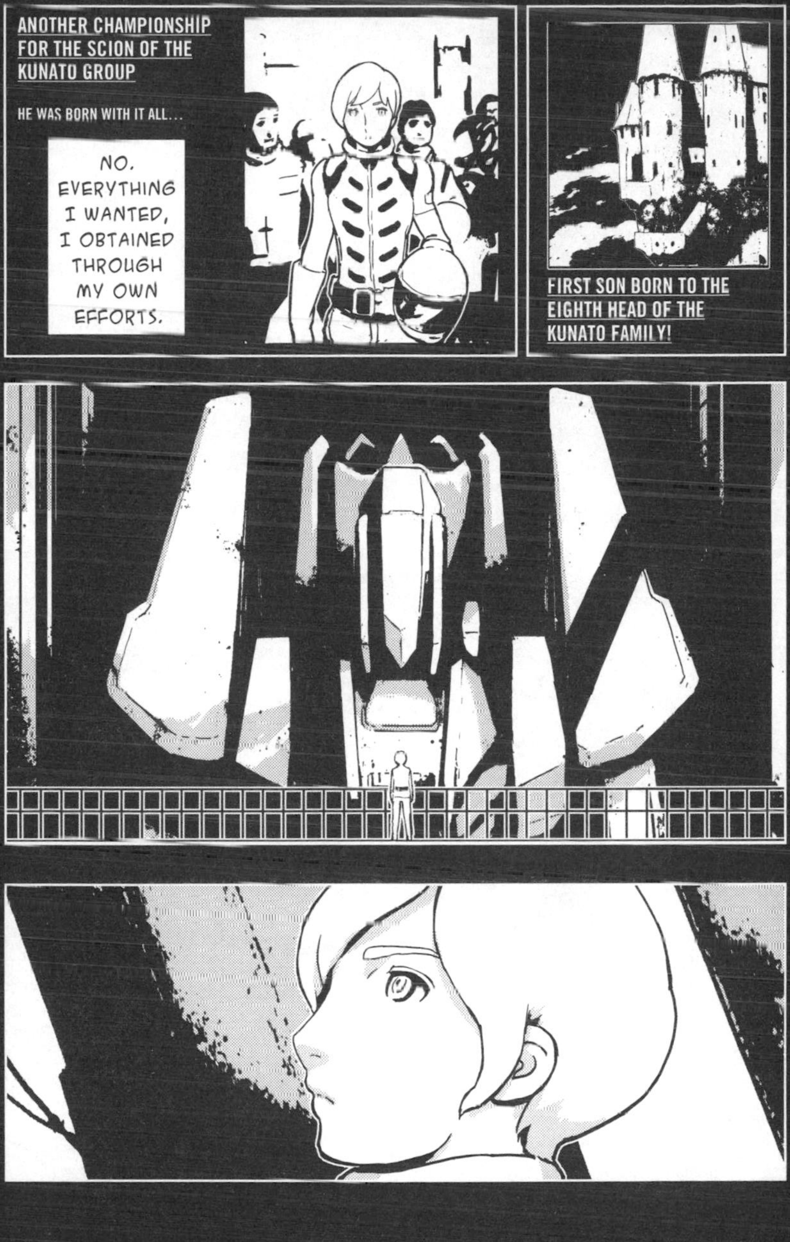
FIRST SON BORN TO THE EIGHTH HEAD OF THE KUNATO FAMILY!
ANOTHER CHAMPIONSHIP FOR THE SCION OF THE KUNATO GROUP
HE WAS BORN WITH IT ALL...
NO. EVERYTHING I WANTED, I OBTAINED THROUGH MY OWN EFFORTS.

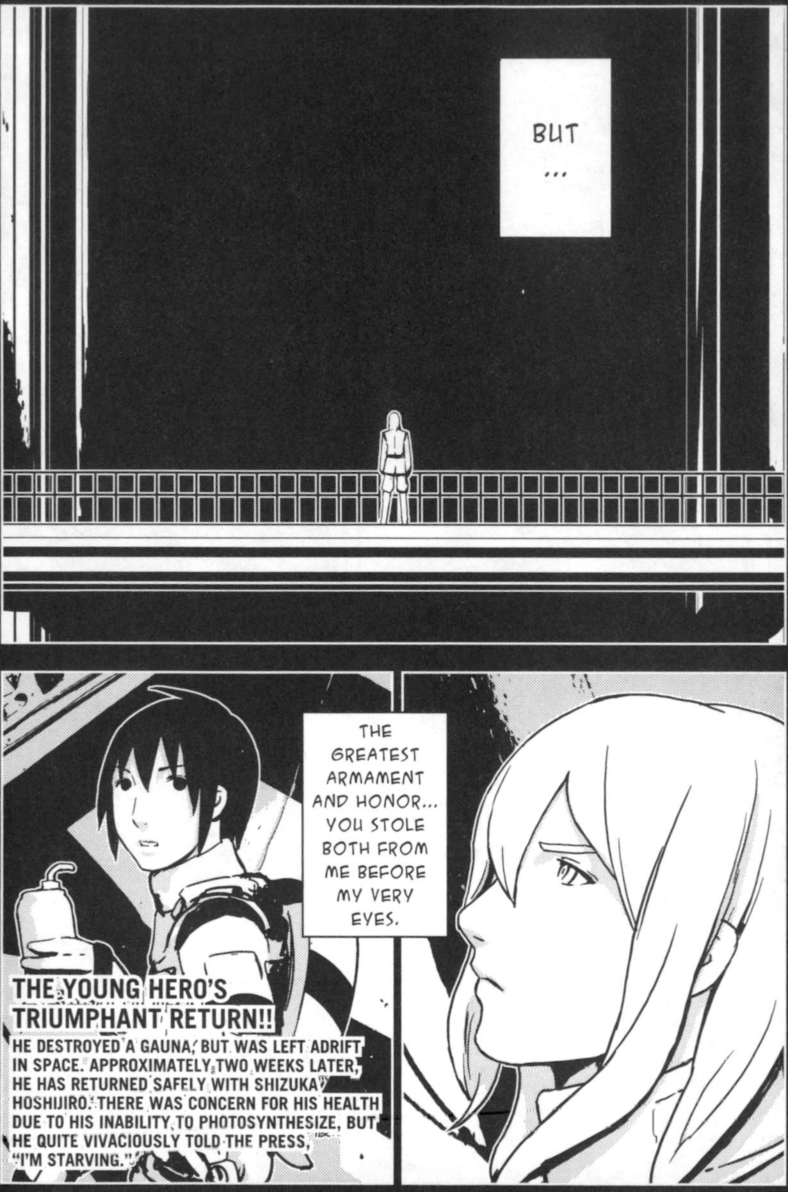
BUT ...
THE GREATEST ARMAMENT AND HONOR... YOU STOLE BOTH FROM ME BEFORE MY VERY EYES.
THE YOUNG HERO'S TRIUMPHANT RETURN!!
HE DESTROYED A GAUNA, BUT WAS LEFT ADRIFT IN SPACE. APPROXIMATELY TWO WEEKS LATER, HE HAS RETURNED SAFELY WITH SHIZUKA HOSHIJIRO. THERE WAS CONCERN FOR HIS HEALTH DUE TO HIS INABILITY TO PHOTOSYNTHESIZE, BUT HE QUITE VIVACIOUSLY TOLD THE PRESS, "I'M STARVING."

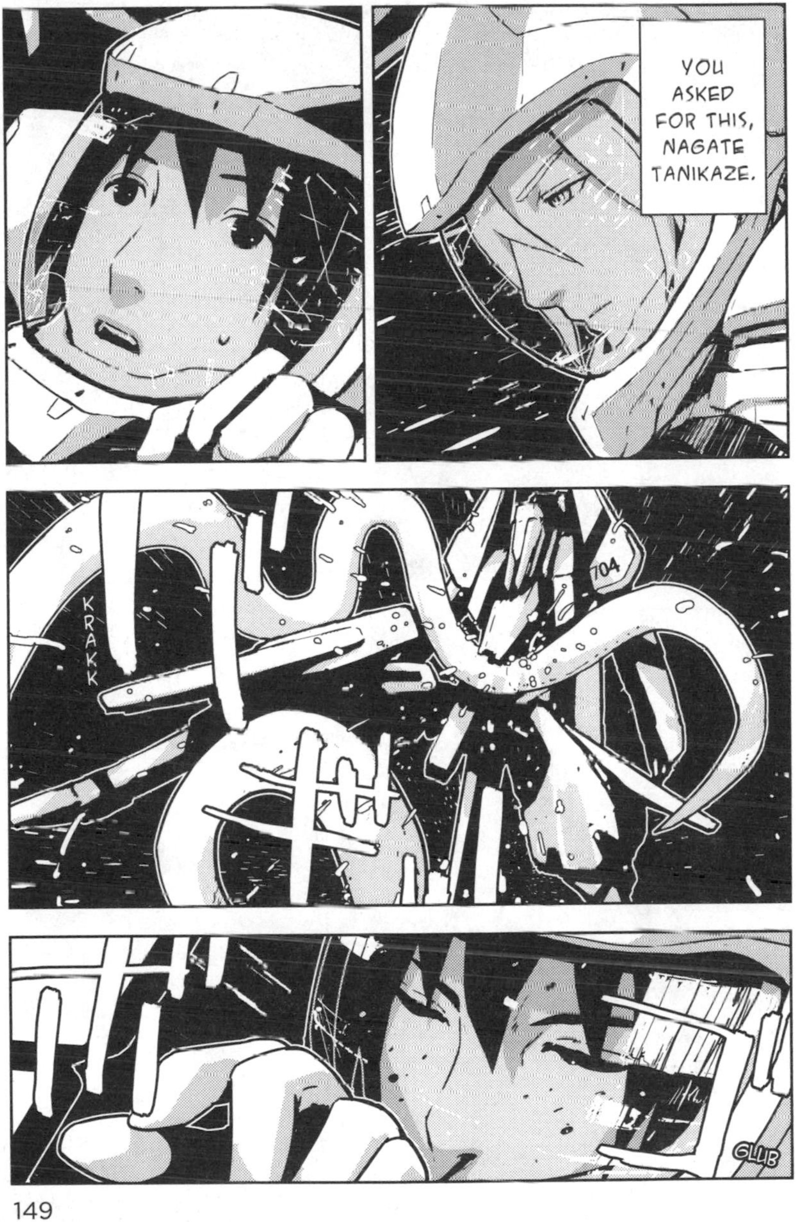
YOU ASKED FOR THIS, NAGATE TANIKAZE.
704
KRAKK
GLUB

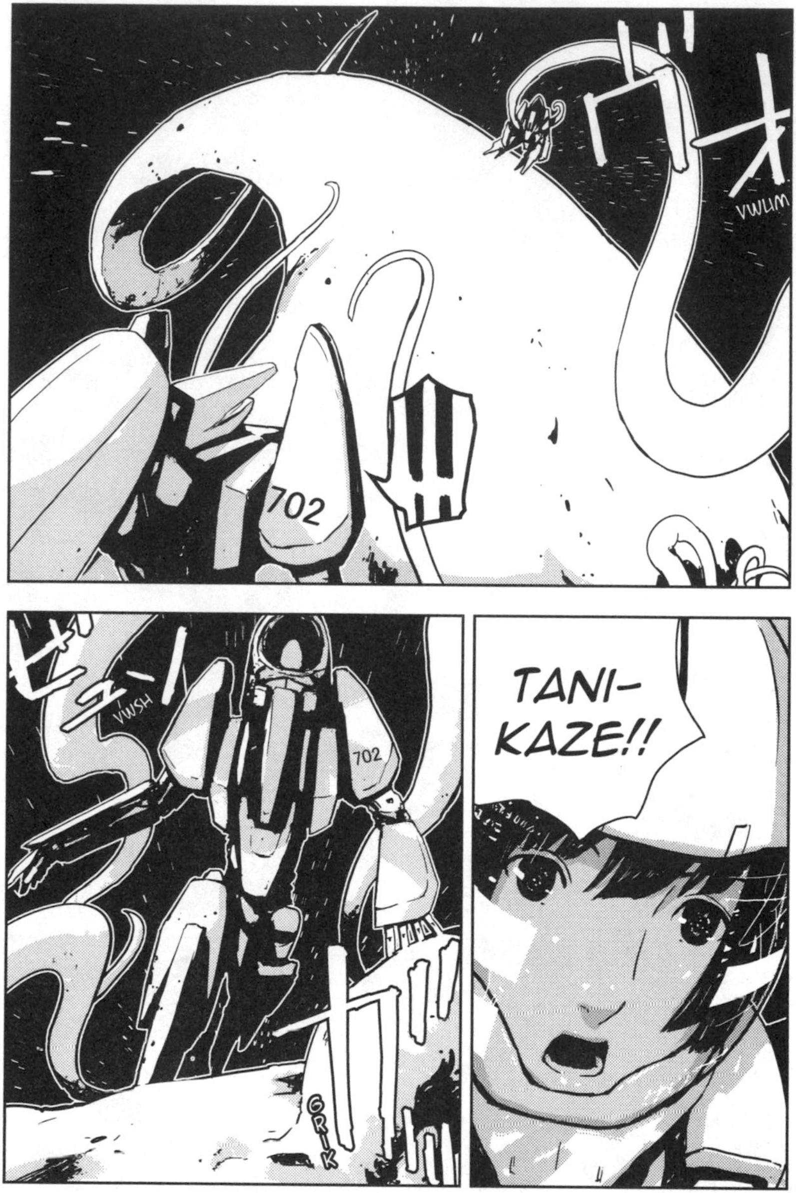
VWUM
702
TANI-
KAZE!!
VWSH
702
GRIK

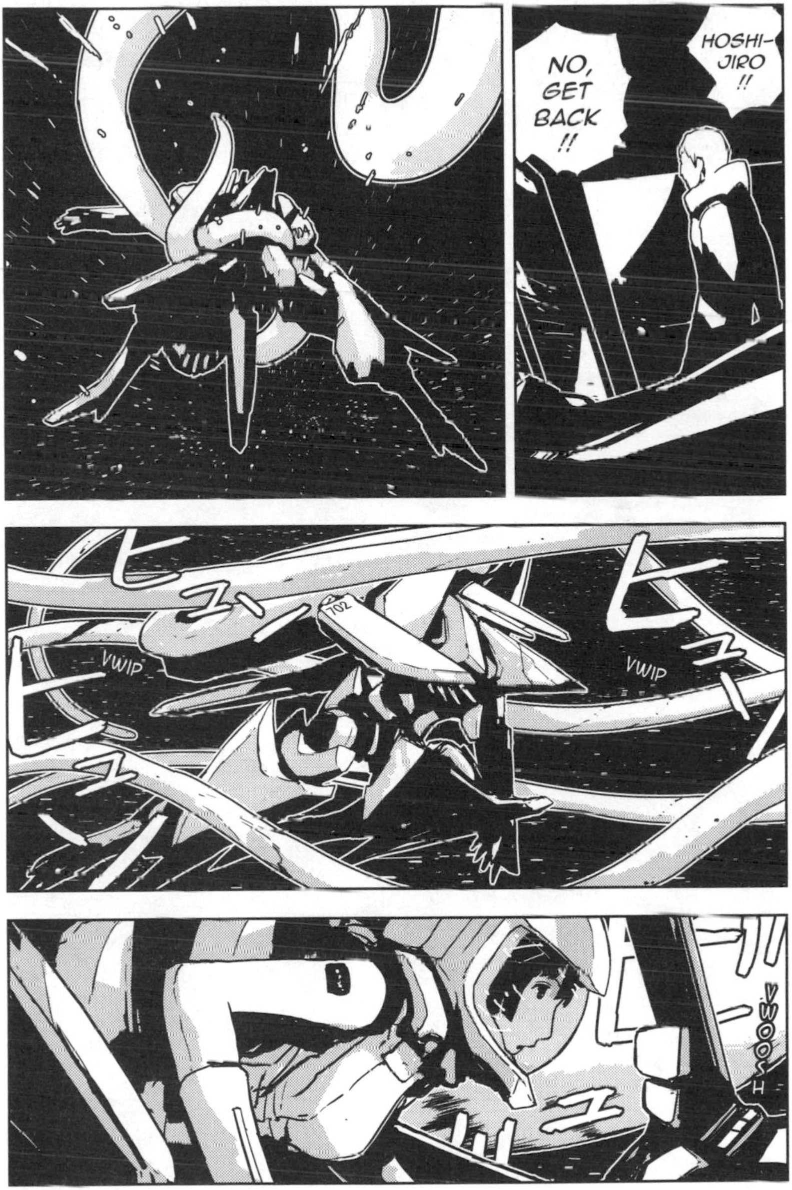
HOSHI-JIRO !!
NO, GET BACK !!
VWIP
VWIP
VWOOSH

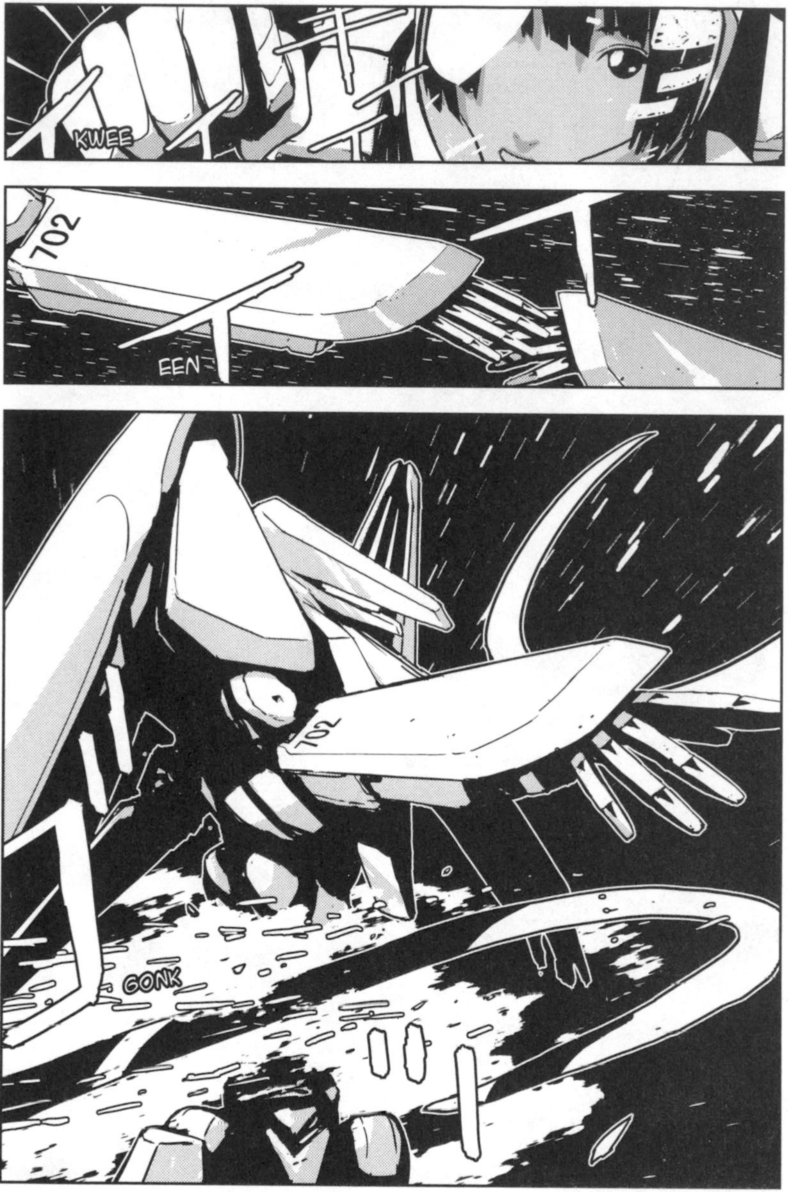
KWEE
702
EEN
702
GONK

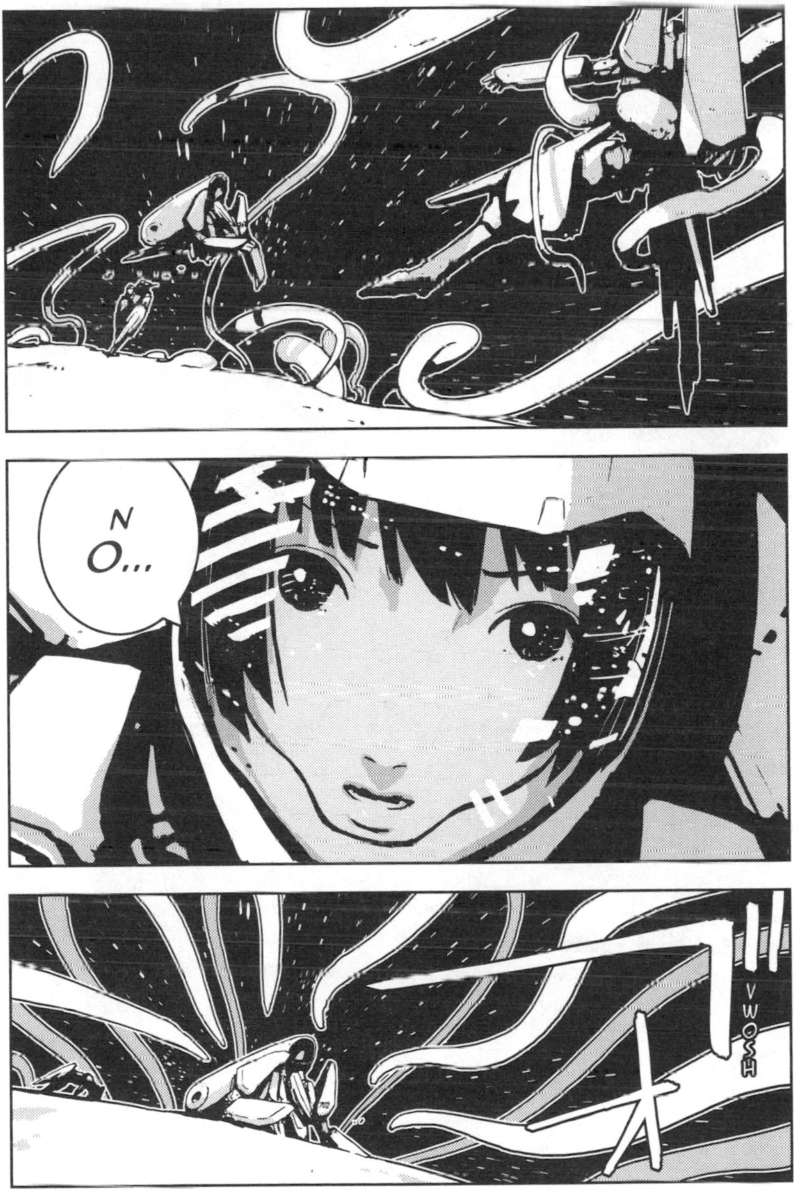
NO...
VWOSH

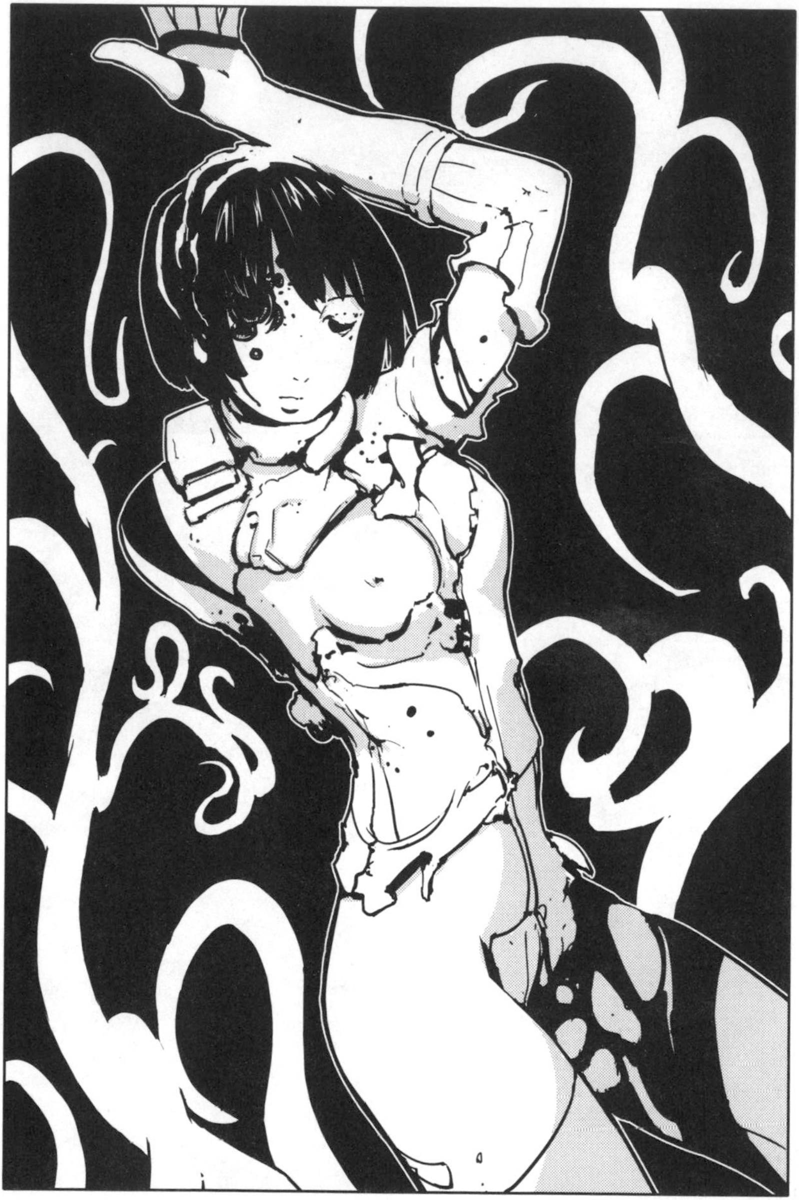

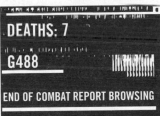

DEATHS: 7

G488

END OF COMBAT REPORT BROWSING

NAGATE TANIKAZE
LOGGED OUT

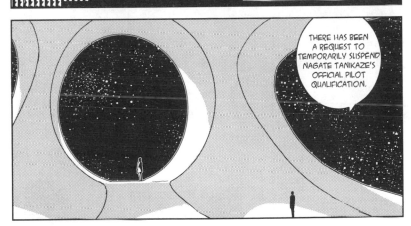

THERE HAS BEEN A REQUEST TO TEMPORARILY SUSPEND NAGATE TANIKAZE'S OFFICIAL PILOT QUALIFICATION.

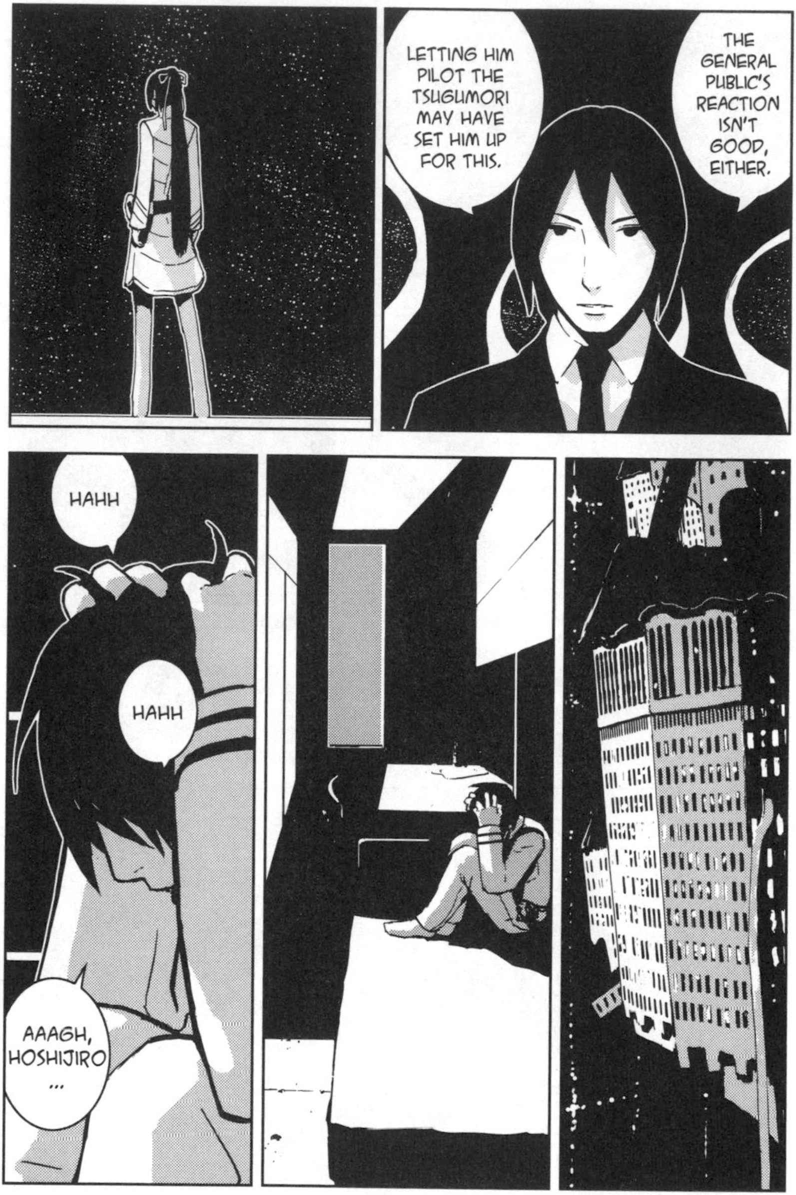
THE GENERAL PUBLIC'S REACTION ISN'T GOOD, EITHER.
LETTING HIM PILOT THE TSUGUMORI MAY HAVE SET HIM UP FOR THIS.
HAHH
HAHH
AAAGH, HOSHIJIRO ...

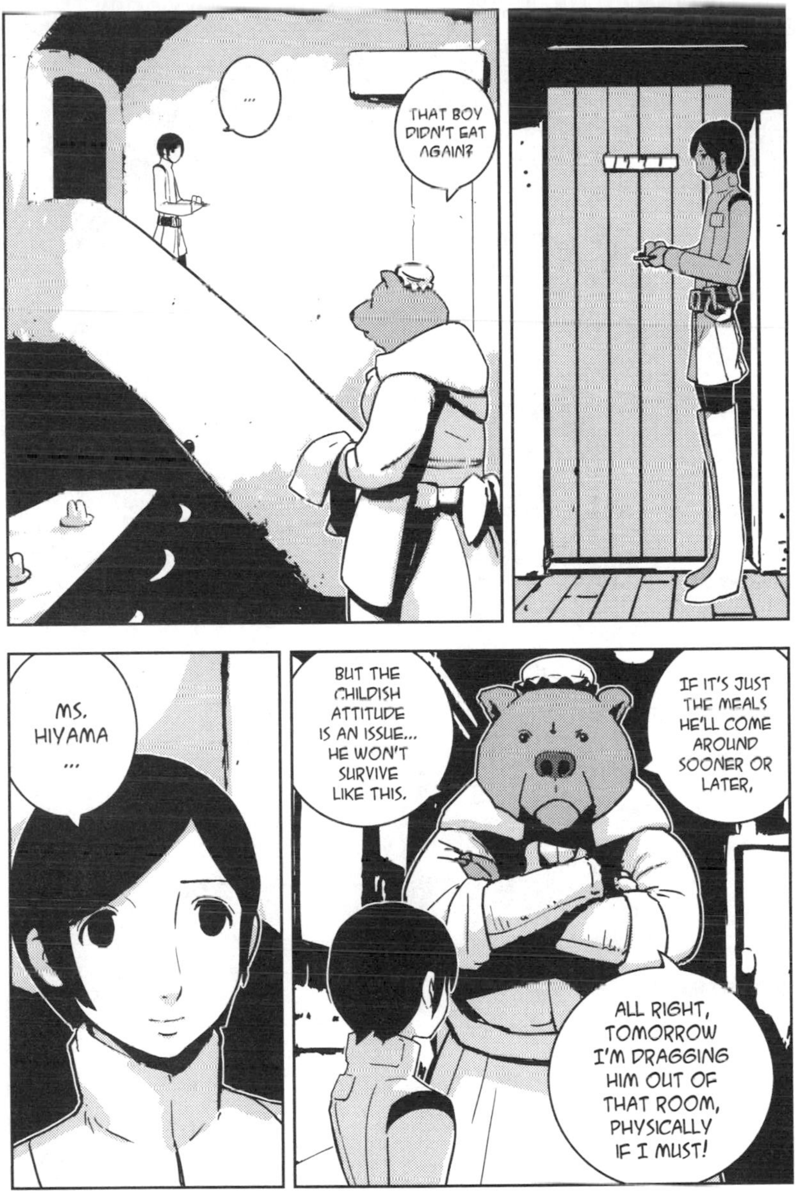
THAT BOY DIDN'T EAT AGAIN?
...
IF IT'S JUST THE MEALS HE'LL COME AROUND SOONER OR LATER,
BUT THE CHILDISH ATTITUDE IS AN ISSUE... HE WON'T SURVIVE LIKE THIS.
ALL RIGHT, TOMORROW I'M DRAGGING HIM OUT OF THAT ROOM, PHYSICALLY IF I MUST!
MS. HIYAMA ...

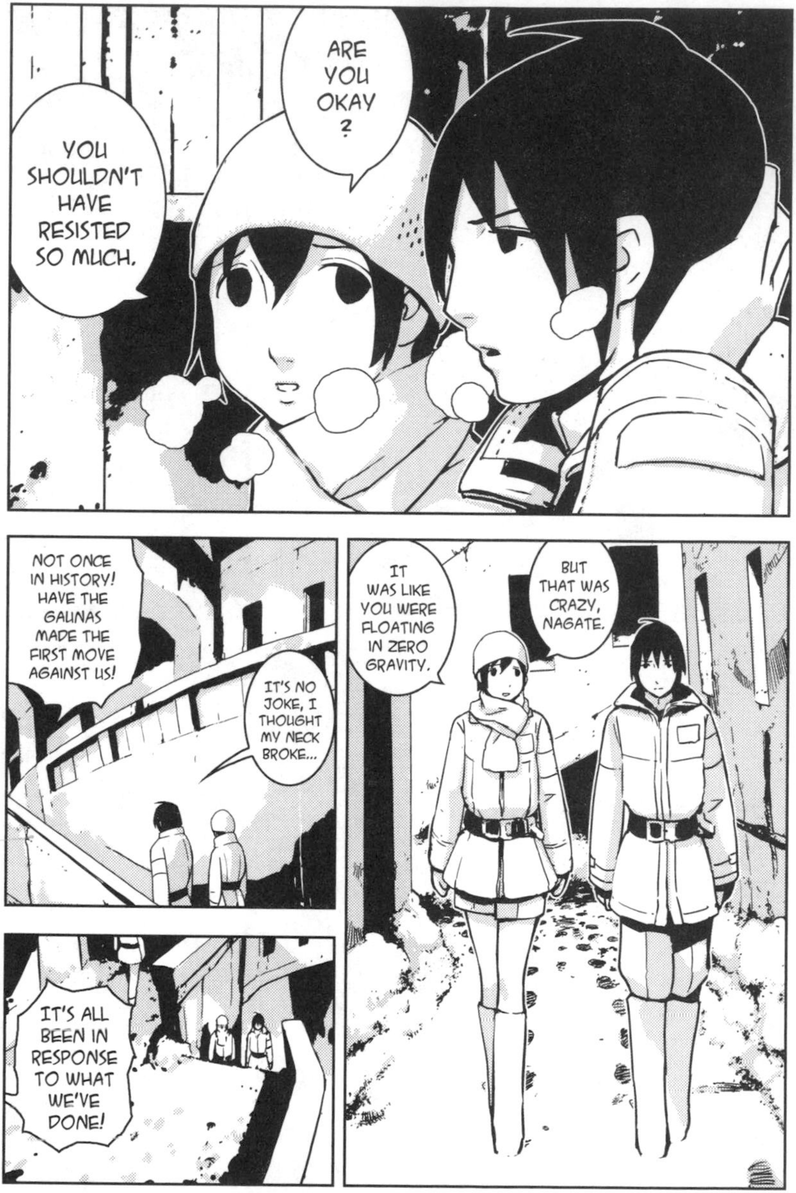
ARE YOU OKAY?
YOU SHOULDN'T HAVE RESISTED SO MUCH.
BUT THAT WAS CRAZY, NAGATE.
IT WAS LIKE YOU WERE FLOATING IN ZERO GRAVITY.
IT'S NO JOKE, I THOUGHT MY NECK BROKE...
NOT ONCE IN HISTORY! HAVE THE GAUNAS MADE THE FIRST MOVE AGAINST US!
IT'S ALL BEEN IN RESPONSE TO WHAT WE'VE DONE!

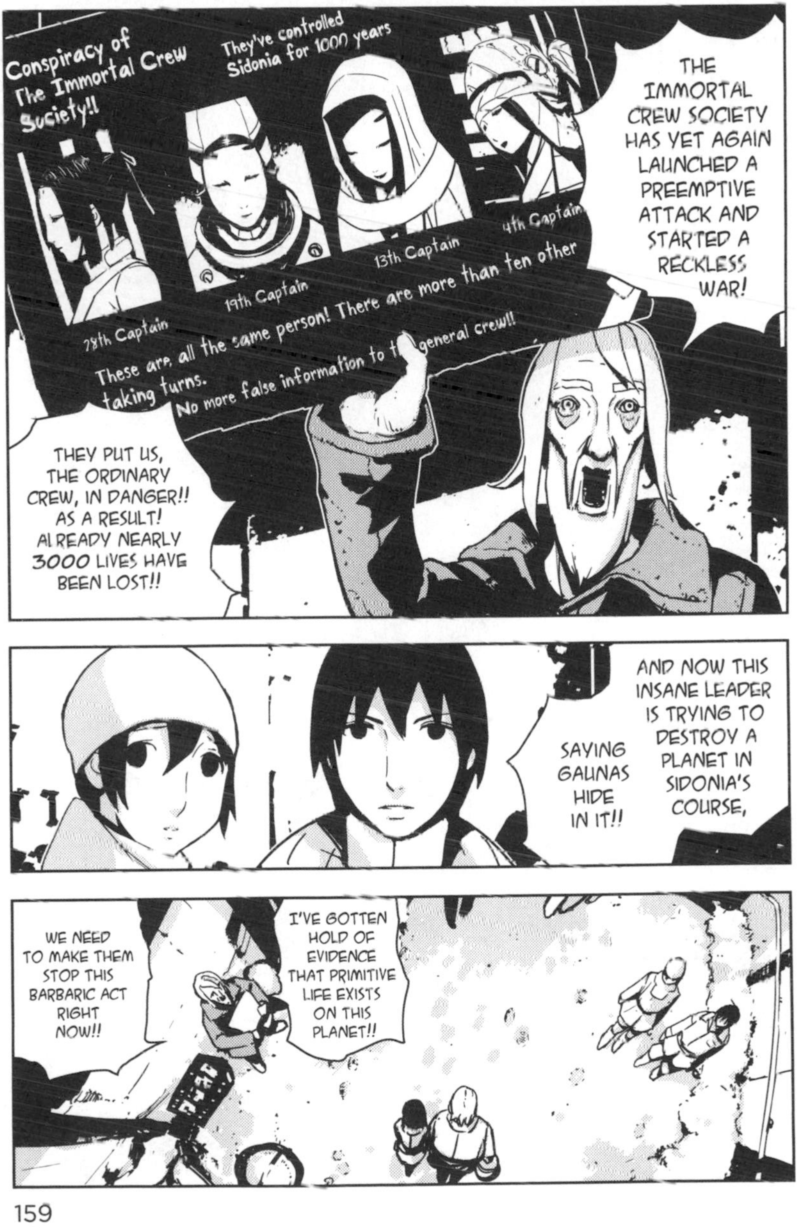
THE IMMORTAL CREW SOCIETY HAS YET AGAIN LAUNCHED A PREEMPTIVE ATTACK AND STARTED A RECKLESS WAR!
They've controlled Sidonia for 1000 years
Conspiracy of The Immortal Crew Society!!
4th Captain
13th Captain
19th Captain
28th Captain
These are all the same person! There are more than ten other taking turns.
No more false information to t general crew!!
THEY PUT US, THE ORDINARY CREW, IN DANGER!! AS A RESULT! ALREADY NEARLY 3000 LIVES HAVE BEEN LOST!!
AND NOW THIS INSANE LEADER IS TRYING TO DESTROY A PLANET IN SIDONIA'S COURSE,
SAYING GAUNAS HIDE IN IT!!
I'VE GOTTEN HOLD OF EVIDENCE THAT PRIMITIVE LIFE EXISTS ON THIS PLANET!!
WE NEED TO MAKE THEM STOP THIS BARBARIC ACT RIGHT NOW!!

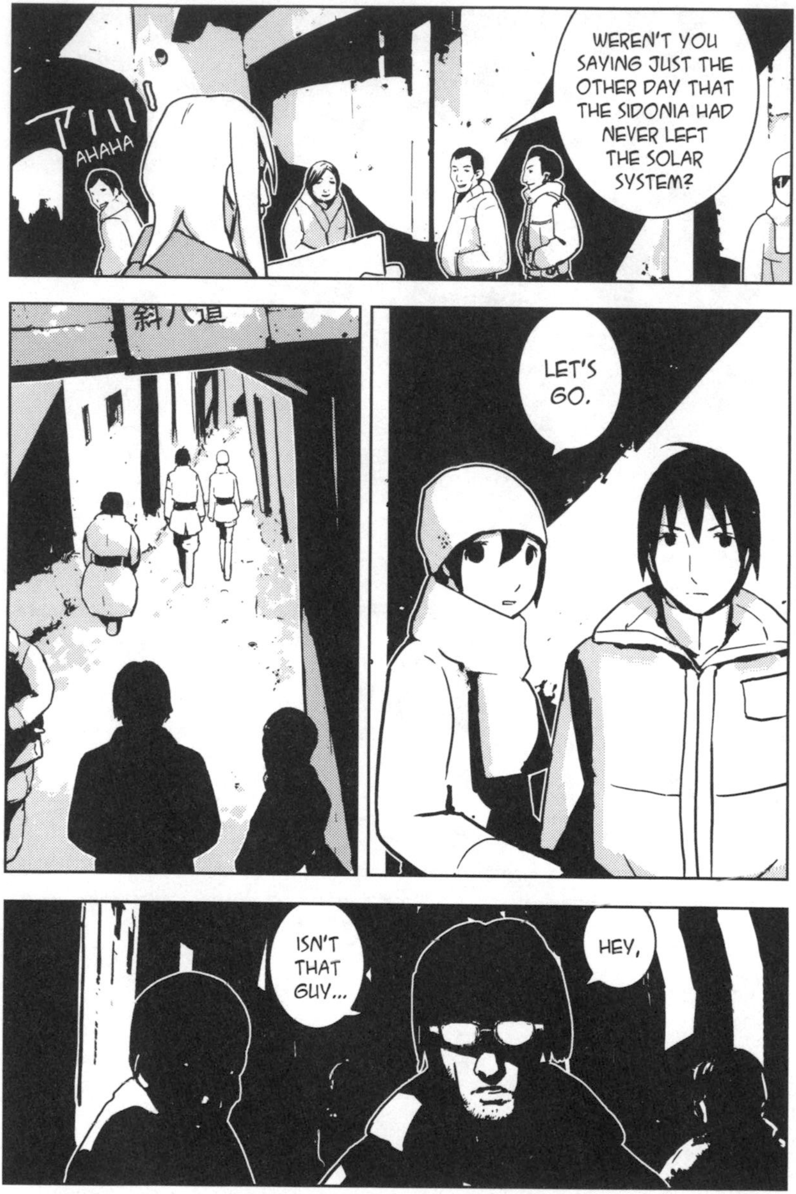
WEREN'T YOU SAYING JUST THE OTHER DAY THAT THE SIDONIA HAD NEVER LEFT THE SOLAR SYSTEM?
アハハ
AHAHA
LET'S GO.
斜八道
HEY,
ISN'T THAT GUY...

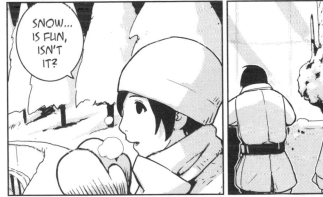

SNOW...
IS FUN,
ISN'T
IT?

YEAH,
LEMME
GO BUY
SOMETHING
WARM TO
DRINK.

...

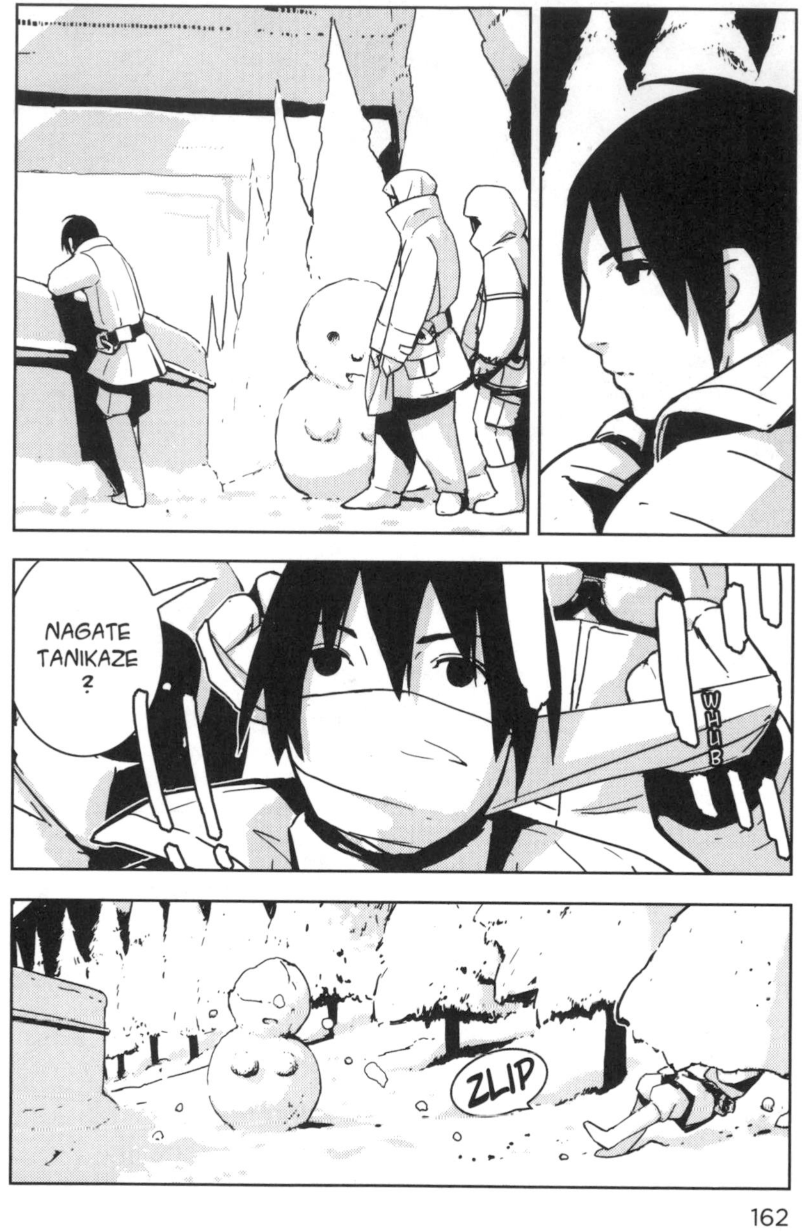
WHUB
NAGATE TANIKAZE ?
ZLIP

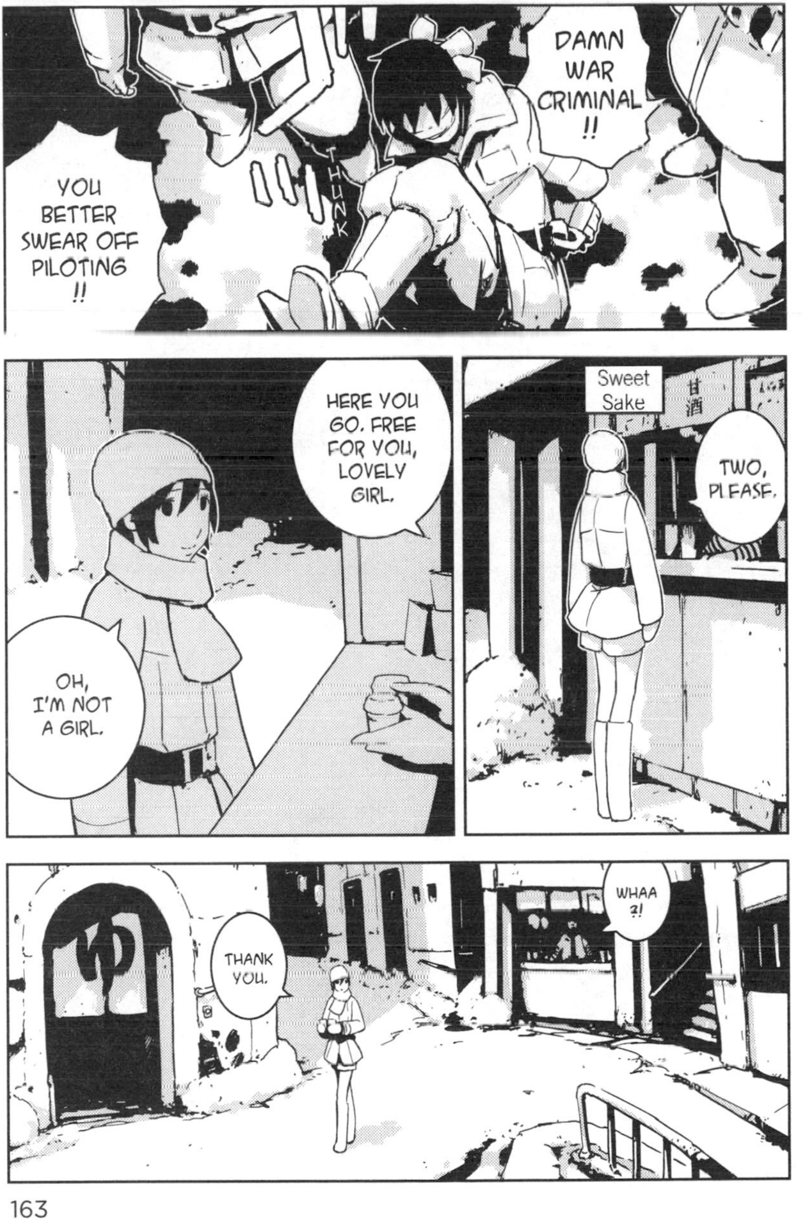
DAMN WAR CRIMINAL !!
THUNK
YOU BETTER SWEAR OFF PILOTING !!
Sweet Sake
甘酒
TWO, PLEASE.
HERE YOU GO. FREE FOR YOU, LOVELY GIRL.
OH, I'M NOT A GIRL.
WHAA ?!
THANK YOU.

ZVSH
ZVSH
NAGATE ?
CLOP

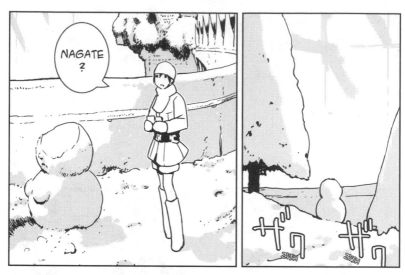

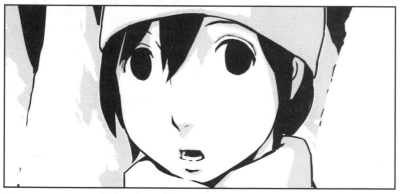

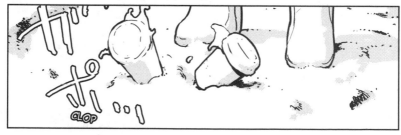

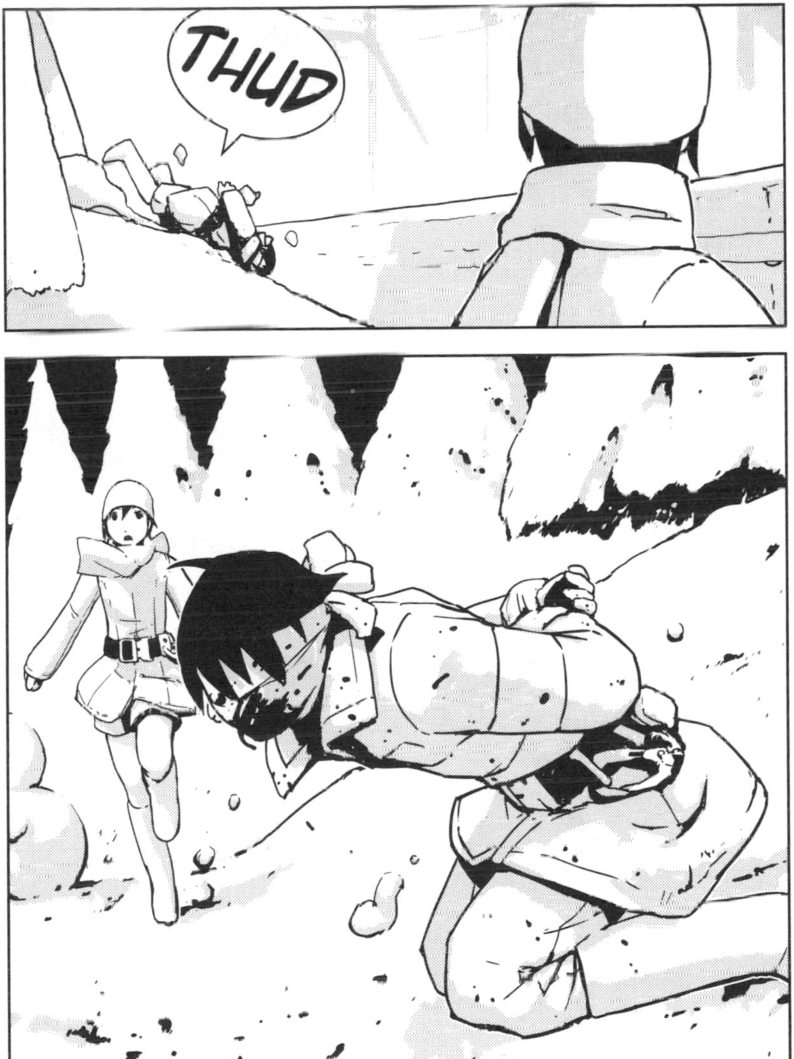
THUD

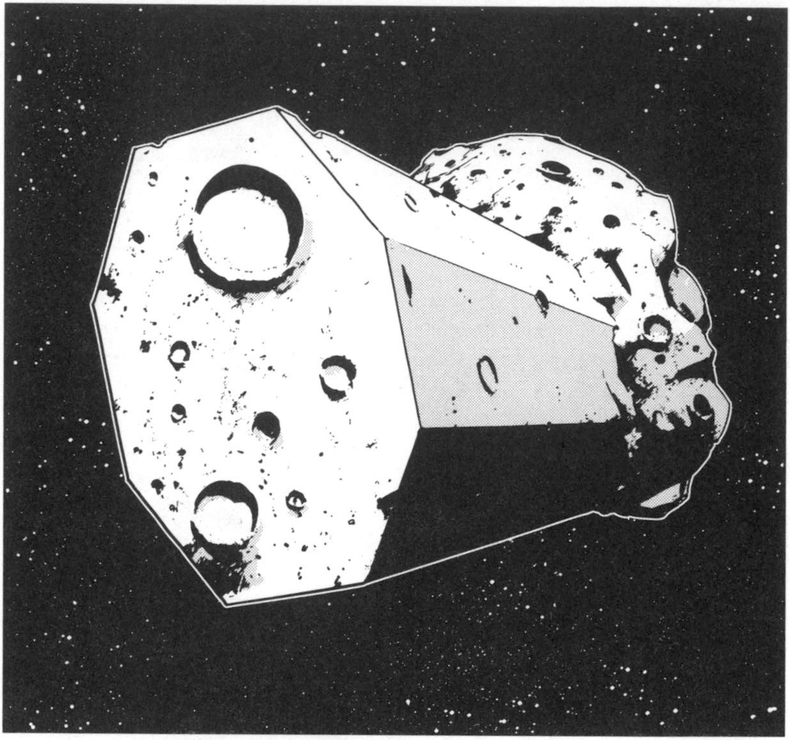

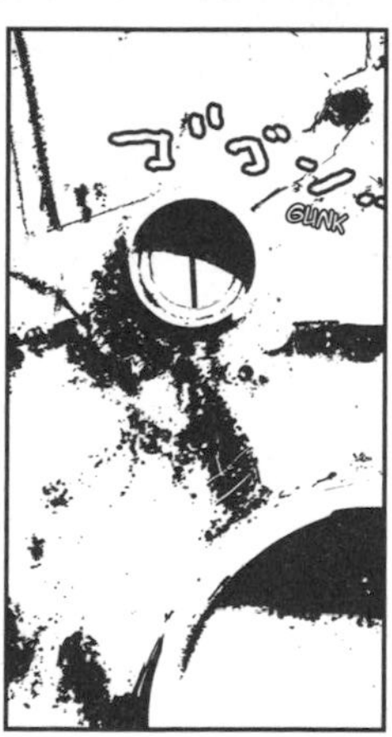
GLINK

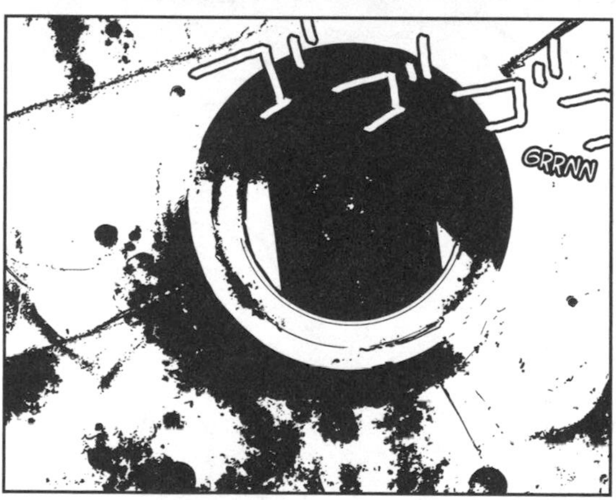
GRRNN

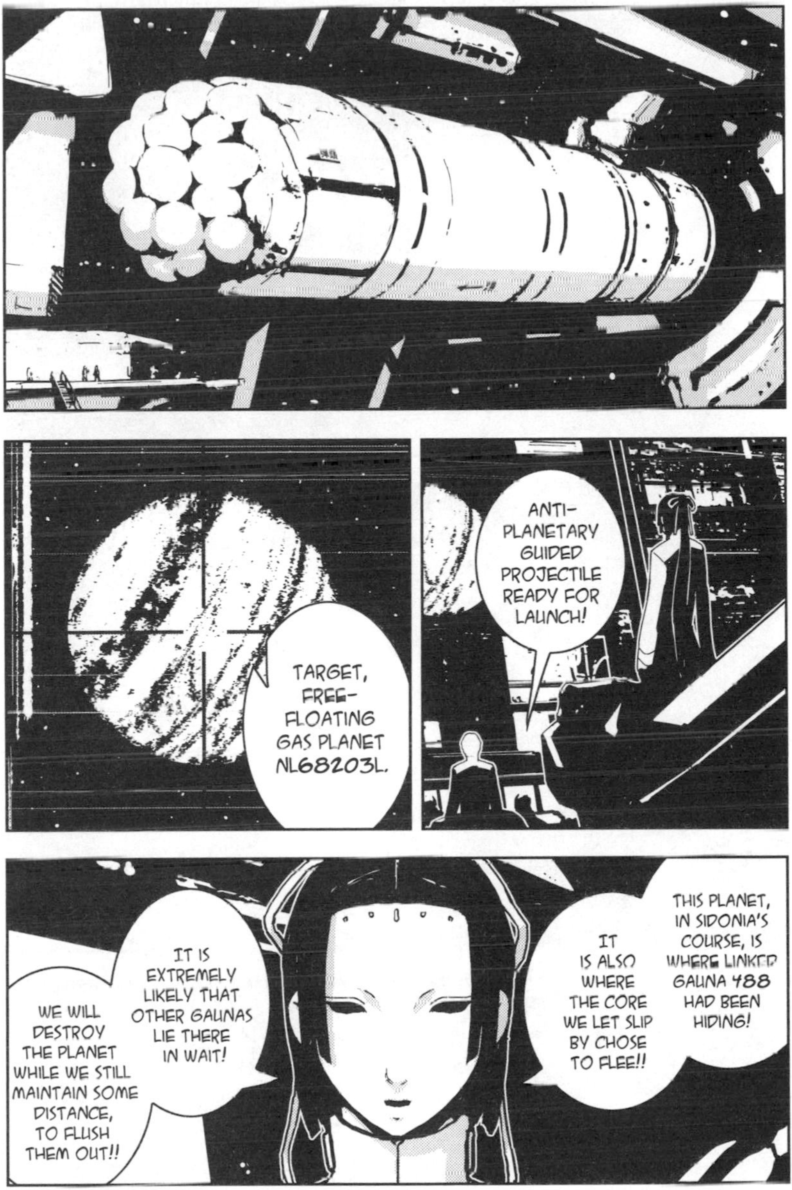
ANTI-PLANETARY GUIDED PROJECTILE READY FOR LAUNCH!
TARGET, FREE-FLOATING GAS PLANET NL68203L.
THIS PLANET, IN SIDONIA'S COURSE, IS WHERE LINKED GAUNA 488 HAD BEEN HIDING!
IT IS ALSO WHERE THE CORE WE LET SLIP BY CHOSE TO FLEE!!
IT IS EXTREMELY LIKELY THAT OTHER GAUNAS LIE THERE IN WAIT!
WE WILL DESTROY THE PLANET WHILE WE STILL MAINTAIN SOME DISTANCE, TO FLUSH THEM OUT!!

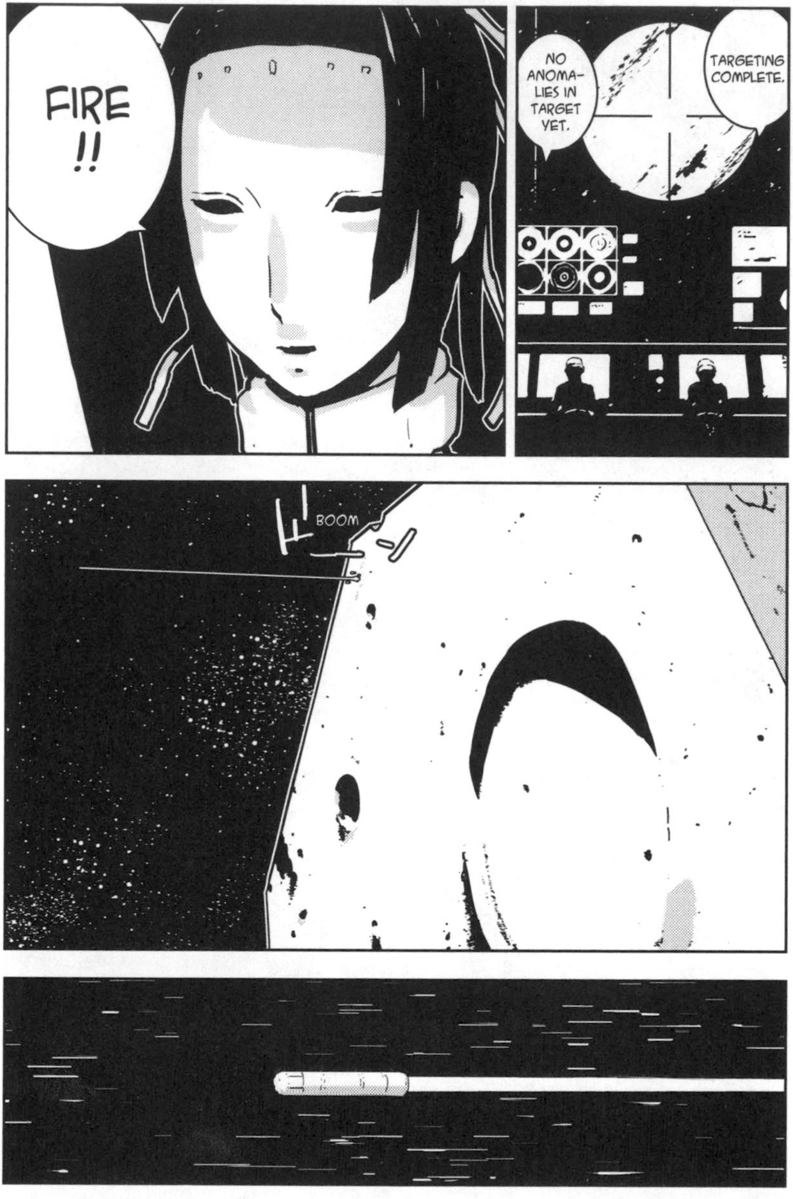
TARGETING COMPLETE.
NO ANOMALIES IN TARGET YET.
FIRE !!
BOOM

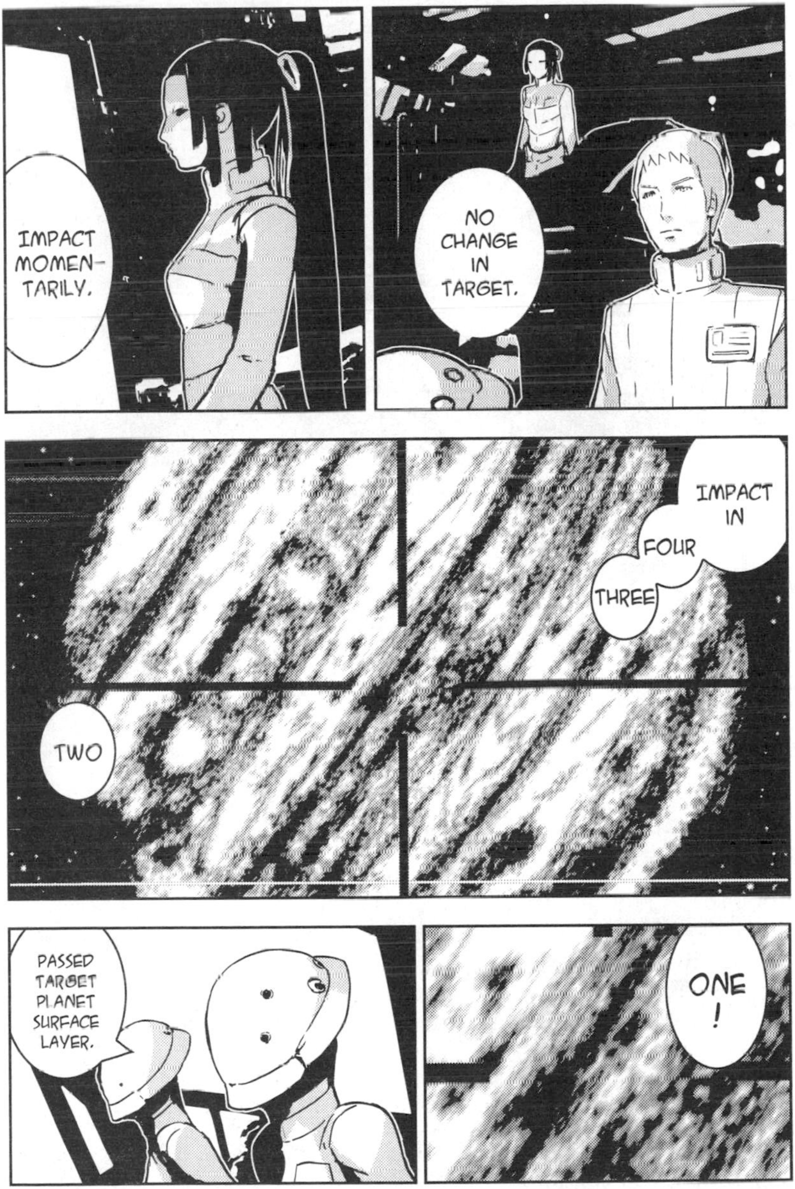
NO CHANGE IN TARGET.
IMPACT MOMEN-TARILY.
IMPACT IN
FOUR
THREE
TWO
ONE!
PASSED TARGET PLANET SURFACE LAYER.

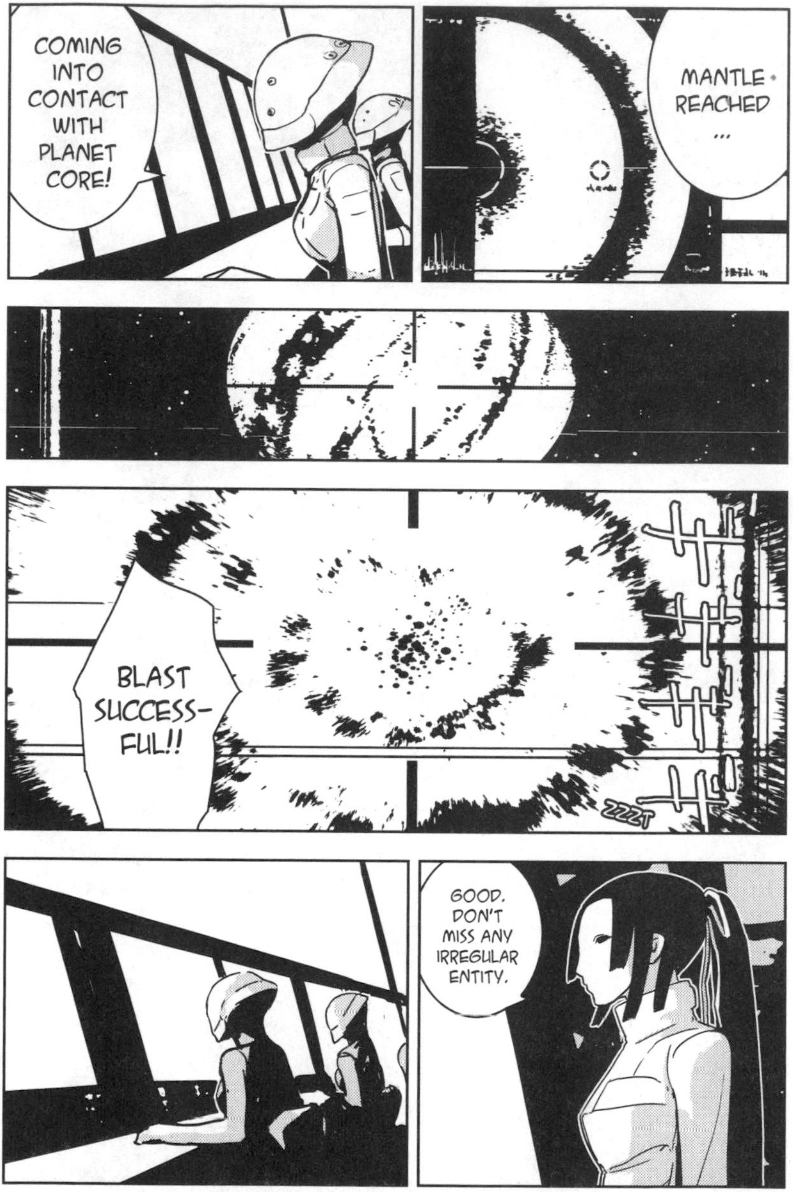
MANTLE REACHED ...
COMING INTO CONTACT WITH PLANET CORE!
BLAST SUCCESSFUL!!
ZZZT
GOOD. DON'T MISS ANY IRREGULAR ENTITY.

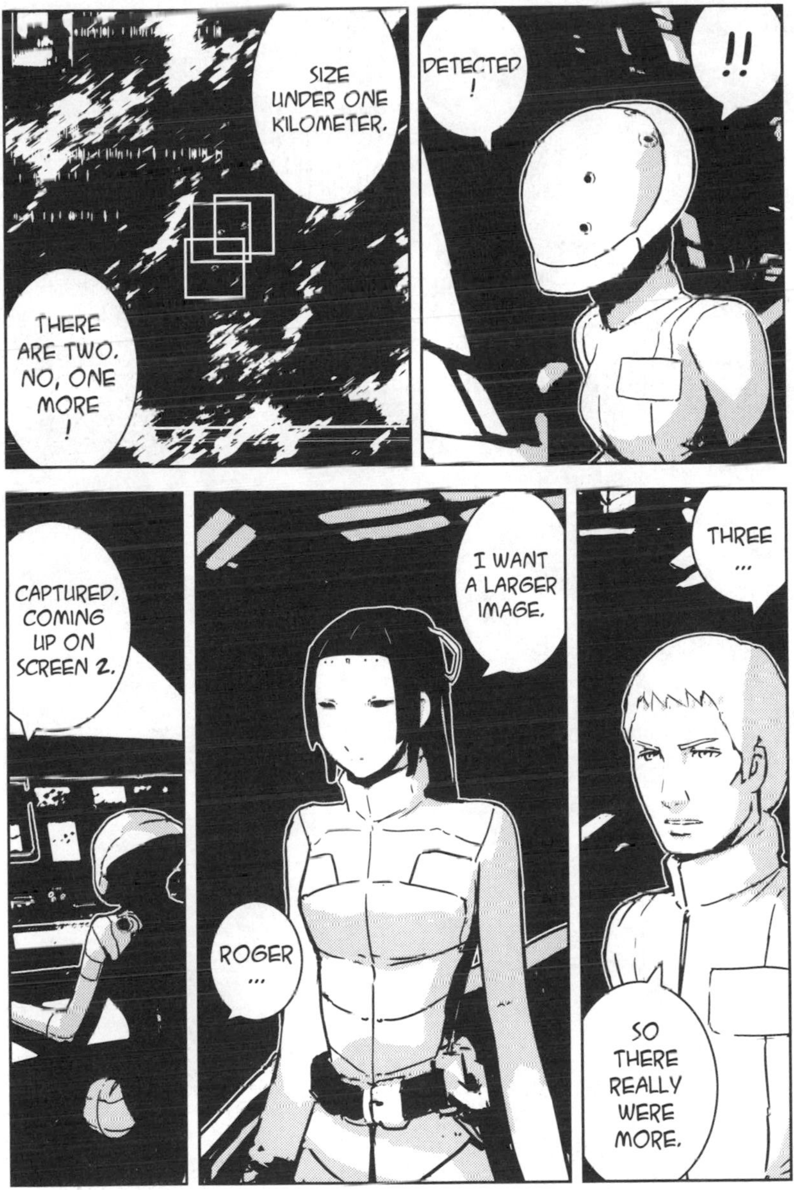
!!
DETECTED !
SIZE UNDER ONE KILOMETER.
THERE ARE TWO. NO, ONE MORE !
THREE ...
SO THERE REALLY WERE MORE.
I WANT A LARGER IMAGE.
ROGER ...
CAPTURED. COMING UP ON SCREEN 2.

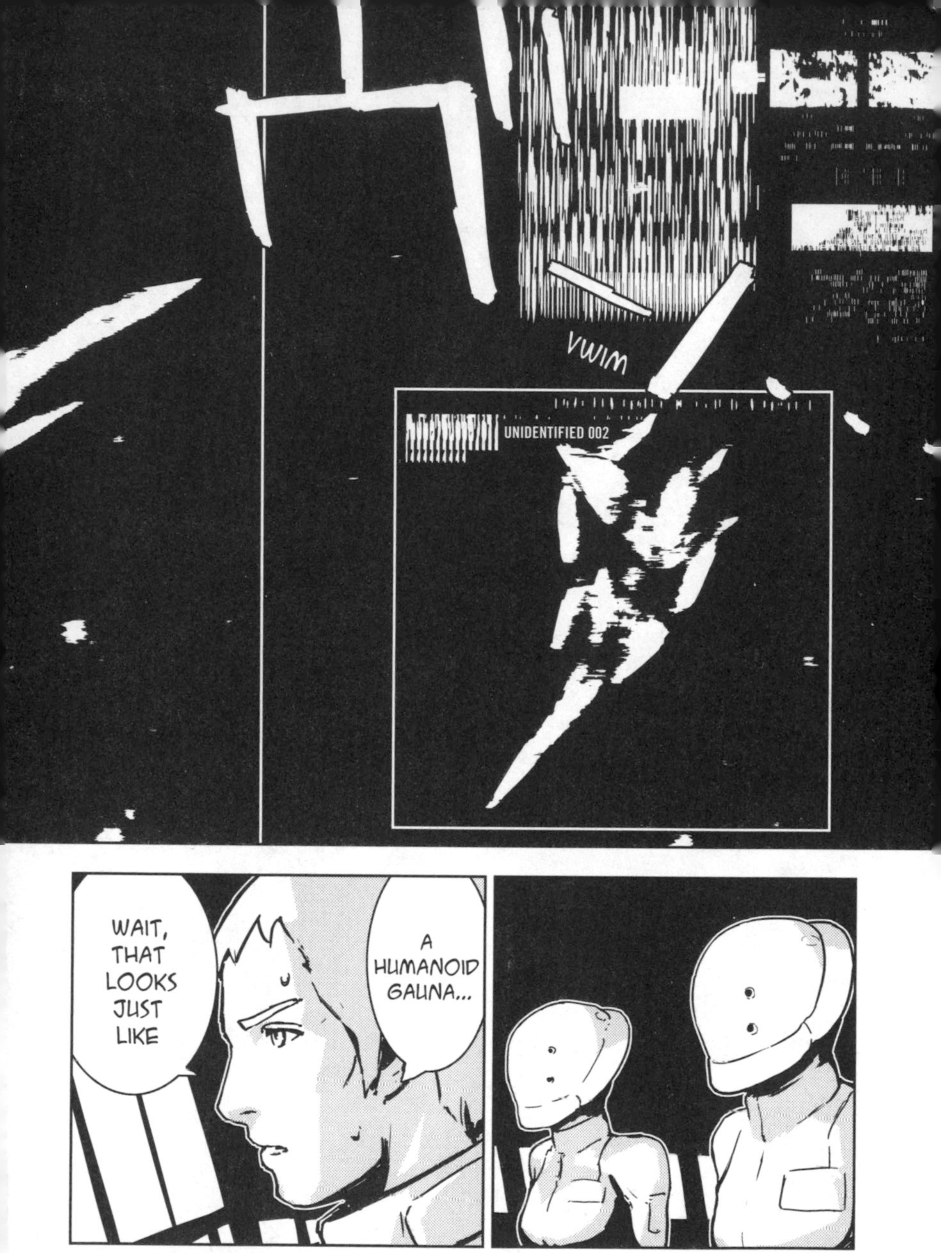
VWIM
UNIDENTIFIED 002
A HUMANOID GAUNA...
WAIT, THAT LOOKS JUST LIKE

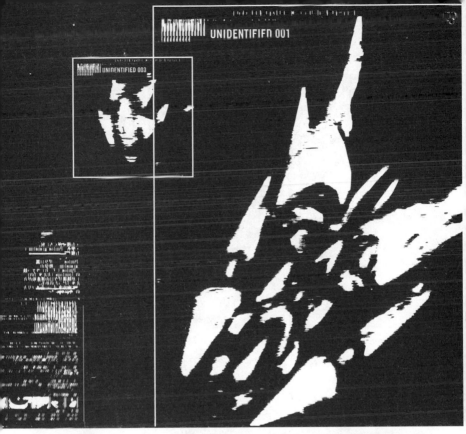

UNIDENTIFIED 001

UNIDENTIFIED 003

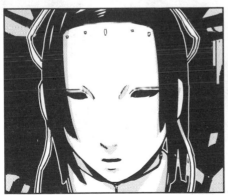

A
GARDE
...

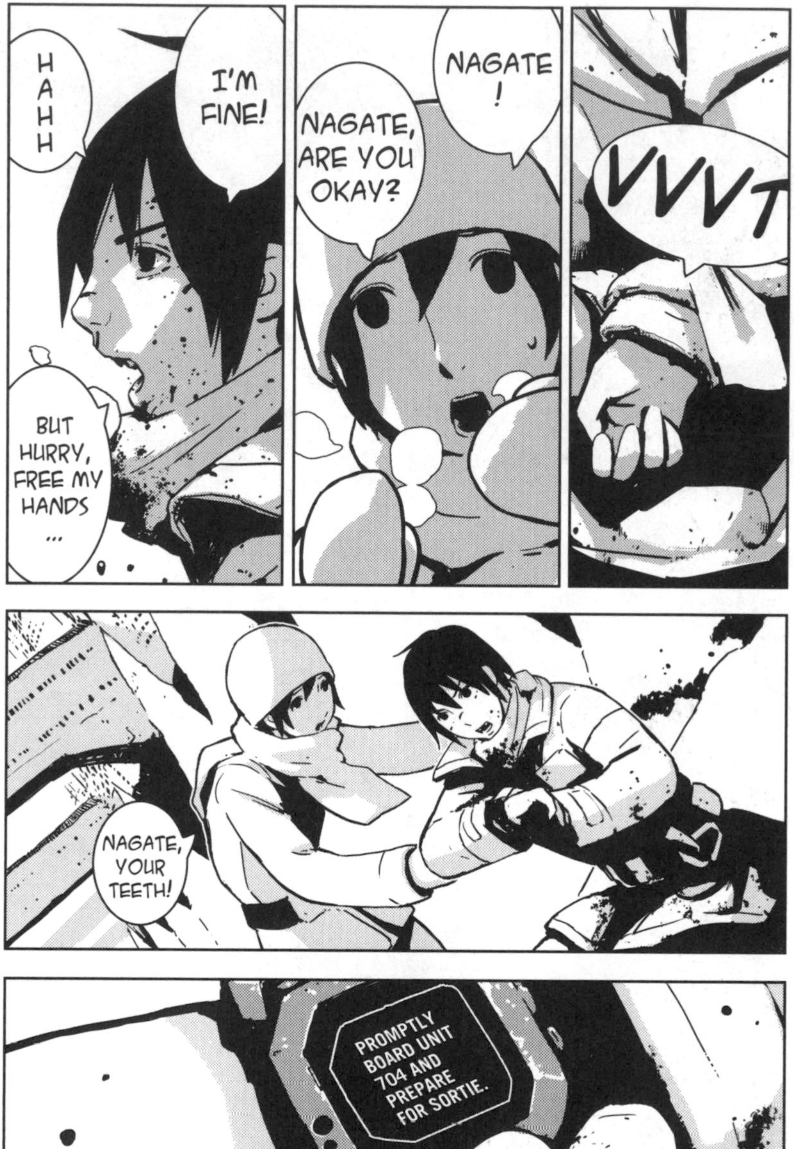
VVVT
NAGATE!
NAGATE, ARE YOU OKAY?
I'M FINE!
HAHH
BUT HURRY, FREE MY HANDS ...
NAGATE, YOUR TEETH!
PROMPTLY BOARD UNIT 704 AND PREPARE FOR SORTIE.

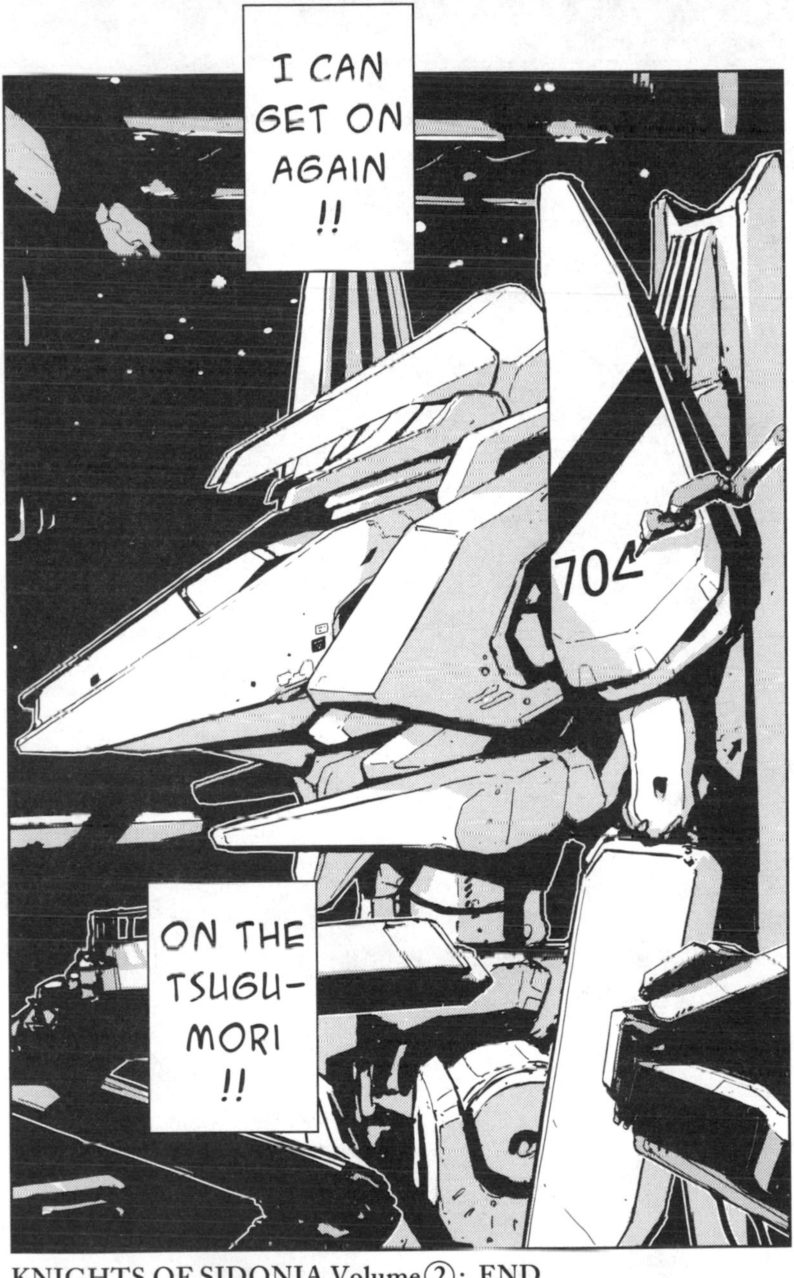
I CAN GET ON AGAIN !!
ON THE TSUGU-MORI !!
704

KNIGHTS OF SIDONIA Volume②: END

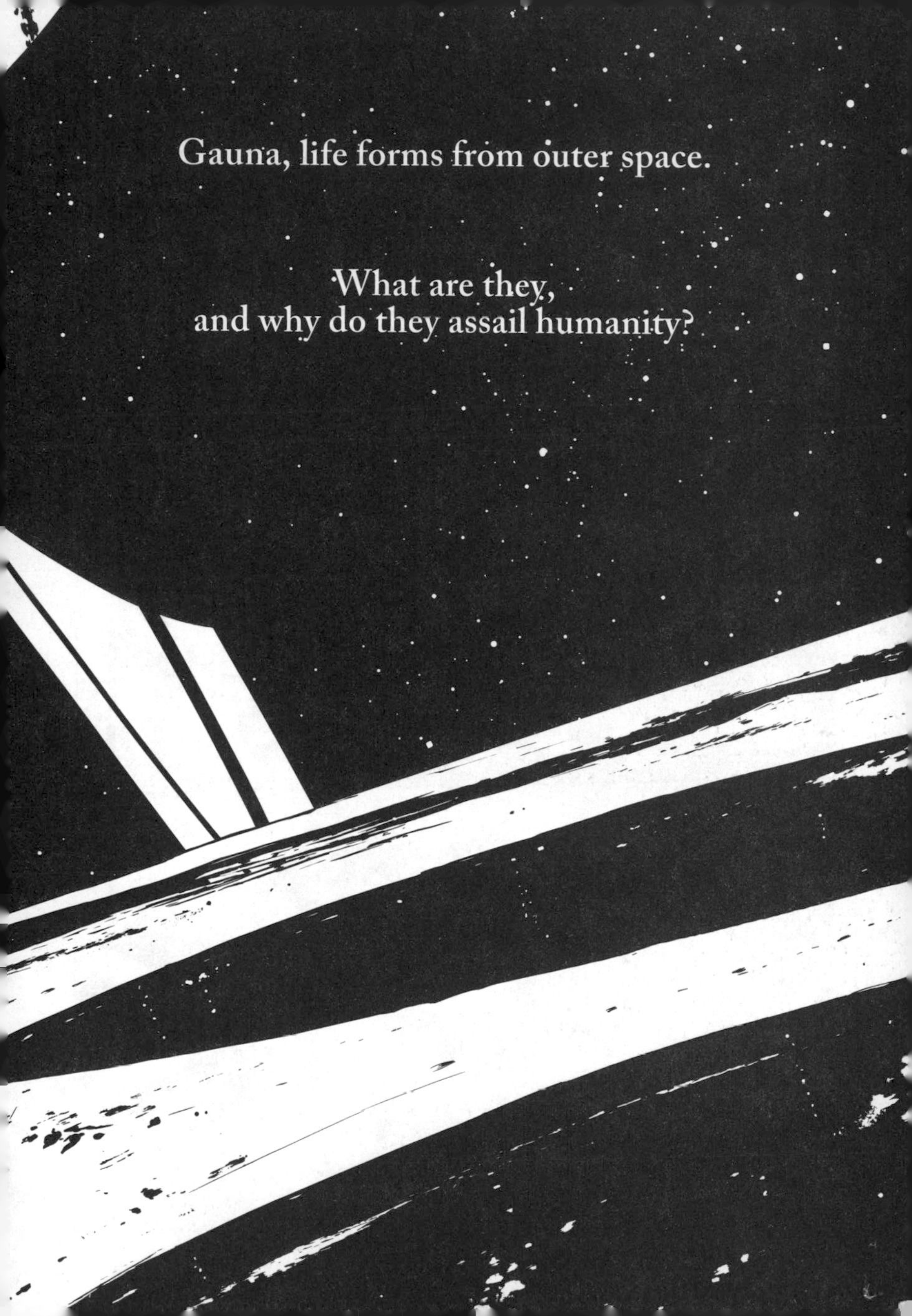
Gauna, life forms from outer space.
What are they, and why do they assail humanity?

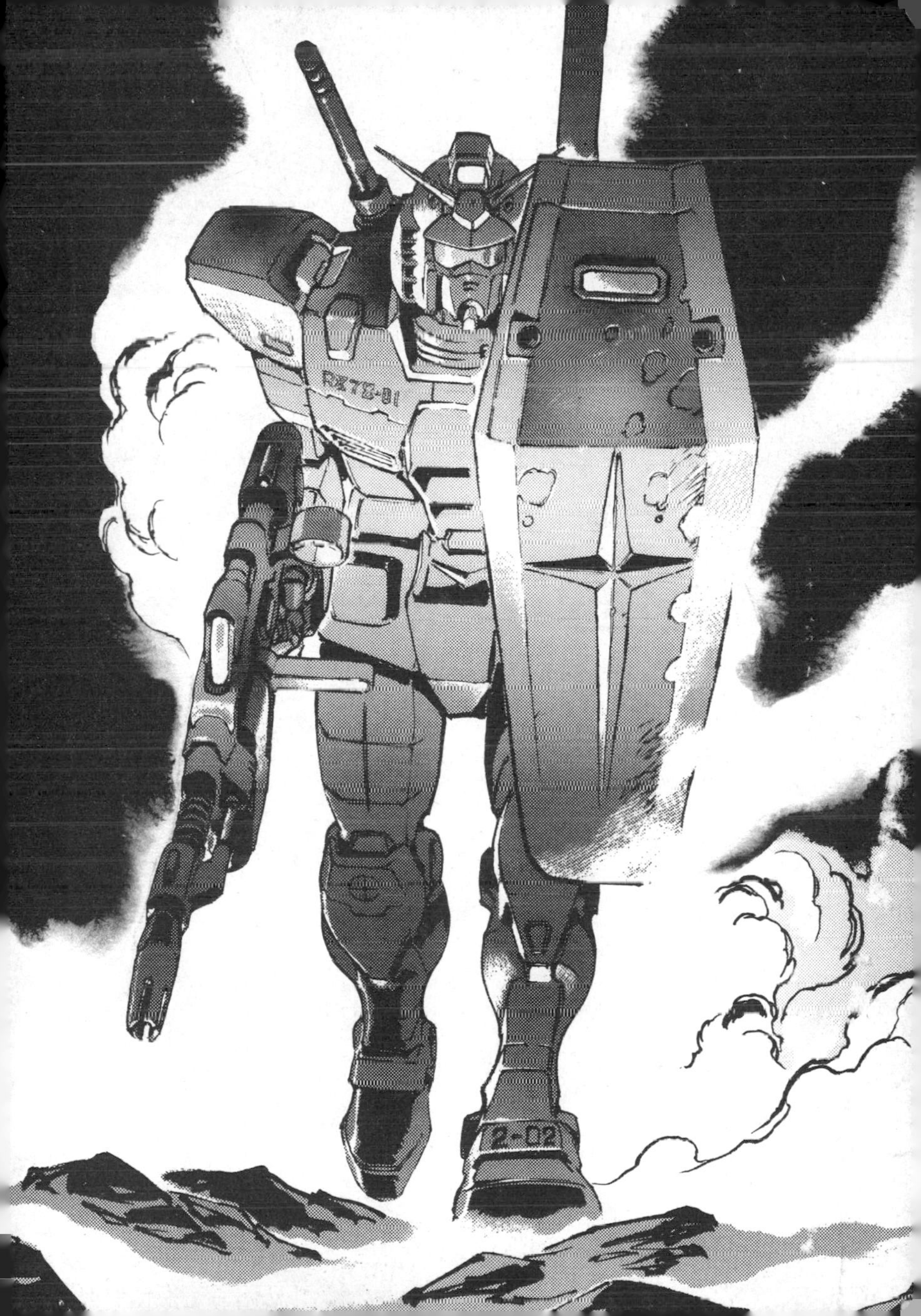
RX78-01
2-02

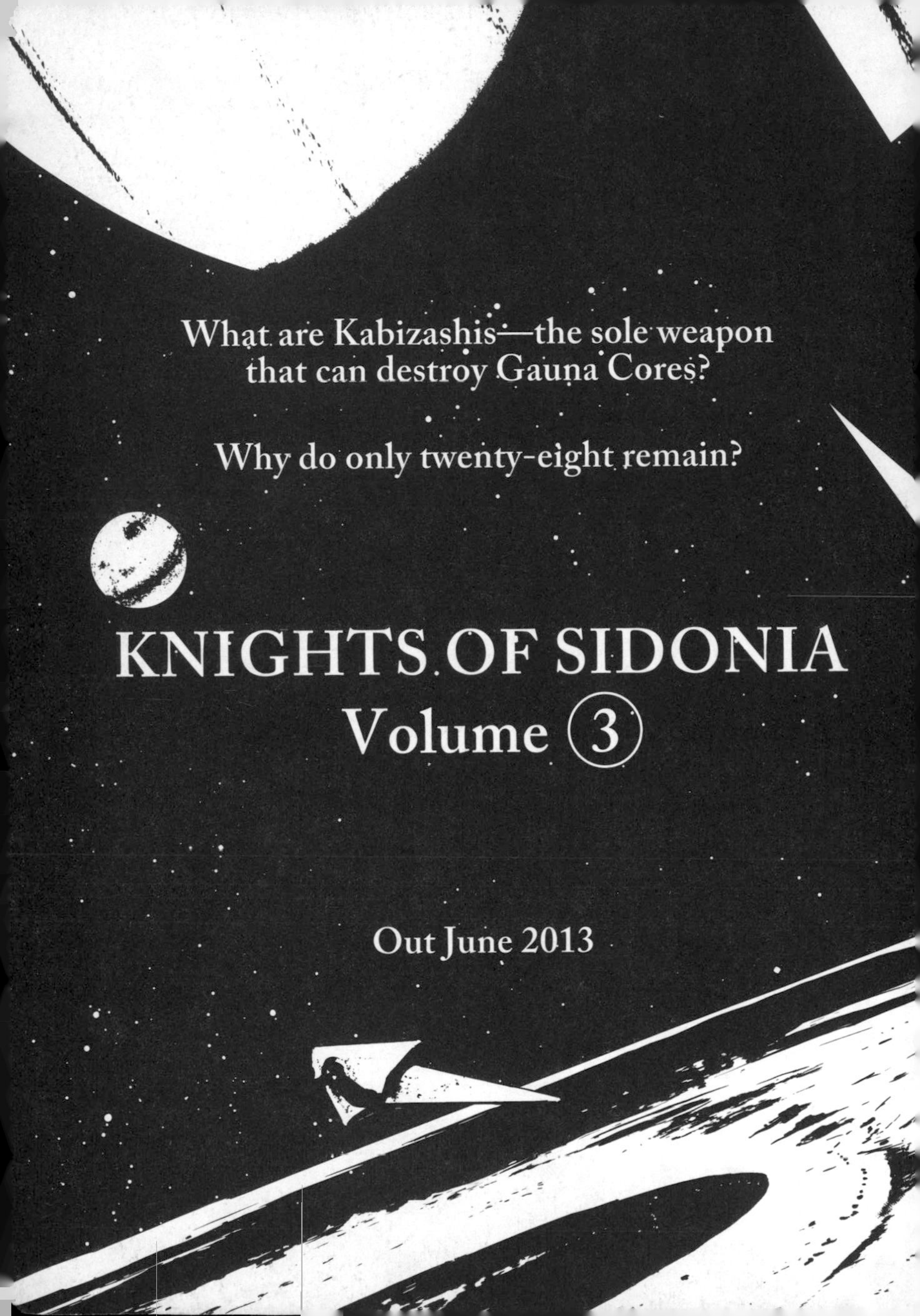
What are Kabizashis—the sole weapon that can destroy Gauna Cores?
Why do only twenty-eight remain?
KNIGHTS OF SIDONIA
Volume ③
Out June 2013